THE SECRET LIVES OF PLANETS

THE SECRET LIVES OF PLANETS

ORDER, CHAOS, AND UNIQUENESS IN THE SOLAR SYSTEM

PAUL MURDIN

PEGASUS BOOKS
NEW YORK LONDON

THE SECRET LIVES OF PLANETS

Pegasus Books, Ltd.
148 West 37th Street, 13th Floor
New York, NY 10018

First Pegasus Books hardcover edition October 2020

ISBN: 978-1-64313-336-2

10 9 8 7 6 5 4 3 2 1

Printed in the United States of America
Distributed by Simon & Schuster
www.pegasusbooks.com

To the engineers and scientists who have shown us far worlds

Contents

CHAPTER 1

Order, chaos and uniqueness in the solar system

If crime fiction is to be believed, English village life is in the main quiet and regular, an ordained series of small and unimportant events punctuated by drama that reveals secrets hidden behind the lace curtains at the windows of outwardly respectable people. The village day has a regular roster of visits by the postman and the electricity-meter reader, the month has a schedule of meetings of the Bridge Club and the church choir, the year has an annual cycle for the Flower and Produce Show and the Nativity Play. But the colonel is then found in his bed, stabbed, it turns out, by a former partner in his shady business dealings in the Far East. The verger is found hanging from the ropes of the church bell, having thus been removed by his ex-wife's lover from the list of beneficiaries of a will. The post-mistress is on the track of the writer of poisoned-pen letters until she is drowned in a well, her bicycle lying nearby on the village green. The quiet life of the village is disrupted and the secrets that lie under its surface are exposed.

Novels like Agatha Christie's are fictionalised versions of real life. We would like to think of life as orderly and structured, but we learn about, and occasionally are participants in, chaotic events like car accidents, illnesses, hurricanes, floods and terrorist attacks.

Likewise, we may have the impression that the solar system is constant and perfectly regular like a clock, or a planetarium instrument. On a short timescale it is. But, seen in a longer perspective, the planets, and their satellites, have had exciting lives, full of drama. As in human lives, some changes in the lives of the planets are evolutionary and gradual, corresponding in us to the natural processes of growing up. Sometimes, they are life-changing, like catastrophic accidents in human lives, that throw a planet into a new trajectory, metaphorically or literally. The effects of the dramatic events leave their traces on the appearance and structure of the planets, and part of the job of planetary science is to infer what happened. 'The present is the key to the past', wrote the nineteenth-century Scottish geologist Archibald Geikie about the Earth. As the Earth, so all the planets.

The vision of the solar system as a clock reached its pinnacle in the eighteenth century. The fundamental geometry of the solar system as a system of planets in orbit around the Sun was surmised by the Polish cleric Nicolaus Copernicus in 1543, and demonstrated to be so by the Italian physicist Galileo Galilei through his discoveries with the telescope in 1610. Empirical laws describing mathematical properties of the planetary orbits, such as the fact that they are ellipses, were established by the German astronomer Johannes Kepler between 1609 and 1619. Drawing together all these discoveries, the mathematician Isaac Newton put forward in 1687 the underlying physical principles of planetary motion in his book known as *Principia*, with its brilliantly simple and exactly formulated notion of a Law of Gravitation.

Newton's model of the solar system held that it was a thoughtful work of mathematics. He asserted in 1726 that 'the wondrous disposition of the Sun, the planets and the comets, can only be the work of an all-powerful and intelligent Being'. According to Newton, God orchestrates the movements of the solar system, and controls them through the Law of Gravitation as the planets progress towards their future.

This model of the Universe developed further in the hands of Newton's successors, notably the French physicist Pierre Simon Laplace. He demonstrated mathematically, from Newtonian principles, that the solar system was stable. The planets orbited in a flat disc around the Sun and they would continue to do so indefinitely. He thought therefore that, once created, the solar system would last in the same form for ever. The solar system was something eternal that developed from its beginnings with inevitability.

Laplace was able to express the certainty of physics with certainty of belief:

> We ought to regard the present state of the Universe as the effect of its antecedent state and as the cause of the state that is to follow. An intelligence knowing all the forces acting in nature at a given instant, as well as the momentary positions of all things in the Universe, would be able to comprehend in one single formula the motions of the largest bodies as well as the lightest atoms in the world, provided that its intellect were sufficiently powerful to subject all data to analysis; to it nothing would be uncertain, the future as well as the past would be present to its eyes.

In an influential book, *Natural Theology or Evidences of the Existence and Attributes of the Deity*, published as the eighteenth century opened, the theologian William Paley described the construction of the planetary system:

> The actuating cause in these [planetary] systems, is an attraction which varies reciprocally as the square of the

distance: that is, at double the distance, has a quarter of the force; at half the distance, four times the strength; and so on . . . So far as these propositions can be made out, we may be said, I think, to prove choice and regulation; choice, out of boundless variety; and regulation, of that which, by its own nature, was, in respect of the property regulated, indifferent and indefinite.

Paley likened the solar system (and human anatomy, and other natural phenomena) to an intricate, well-made watch. He inferred from this that, just as a watch was made in a particular way by a watchmaker, natural phenomena were made by God, the Divine Watchmaker. This is the Teleological Argument for the existence of God (otherwise known as the Argument from Design). In brief, the argument is: natural phenomena work well; they fit together intricately as if designed; there must have been a Designer; the designer is God. Paley reasoned that, if we find a watch lying on the ground,

the inference, we think, is inevitable; that the watch must have had a maker; that there must have existed, at some time and at some place or other, an artificer or artificers, who formed it for the purpose which we find it actually to answer; who comprehended its construction, and designed its use.

It was a reassuring model of the Universe: we live in a harmonious world designed by the Supreme Being. Paley applied this idea to the solar system of planets, but he concentrated also on human anatomy – the human eye looked as if it had been made to a design and God was that designer. The model persists in modern times, and Paley's book is still quoted.

The nineteenth century found an alternative natural theory to account for the structure of the human body in Darwin's Theory of Evolution. In living creatures, the design is only apparent, because natural variations inherited from a parent are passed on to subsequent generations if the variations are favourable to biological success. There is thus a repeated, incremental process by which the structure of a biological organ improves, the better to suit its uses. It only seems as if the organ was designed on purpose. The argument in Paley's book is used nowadays principally to support opposition to Darwin's Theory of Evolution, often in favour of Creationism, the assertion that the Universe, in particular humankind, was created once and for all by God.

In biology, the scientific argument is that living things evolve towards an apparently foreseen design through incremental, hereditable changes that result in improvements in function via natural selection. In physics, the scientific advances of quantum mechanics came about in the twentieth century and cast Paley's expressions of confidence about the functioning of physics, based on Natural Theology, into postmodern doubt. Quantum mechanics explicitly brought into play an Uncertainty Principle: the outcome of a given process in physics is inherently uncertain, and there is no inevitability to the result of a natural physical change, just a range of possibilities, some more favoured than others.

This is most readily apparent in the behaviour of small things – electrons, atoms, quarks, etc. In astronomy, the future of large things – such as the solar system – is also uncertain, due in that case to chaos theory, which was discovered in applications of the theory of gravitation to astronomy. Laplace's claim of Enlightenment certainty, that he could in principle predict everything that will take place in the future using the theory of gravitation, is untrue. There is no certainty in the future, only probability. This is the reverse of what we need from a clock design.

In boasting about what a powerful intelligence could foresee, Laplace was extrapolating from Newton's analysis of two bodies in orbit one around the other: the Sun and a planet, or two stars or two galaxies. The orbits in these cases are indeed determined for all time, ellipses that repeat indefinitely. But, of course, the solar system consists of more than two bodies – there are eight major planets in orbit around the Sun, and innumerable smaller bodies. At some level, it is impossible to ignore the pull of each planet on the others, and the orbits of planets are actually much more complex than the repetitive ellipses of the simple two-body case.

The extension of Newton's theory from two bodies, even by only one more to just three bodies, proved difficult, indeed, intractable. In 1887 the King of Sweden offered a prize for the solution of what came to be known as the Three-Body Problem: what are the orbits of three bodies moving under the influence of their mutual attraction by gravity? The French mathematician Henri Poincaré entered the competition and won because his analysis was the most impressive entry, but he did not find the precise, mathematical solution that was being sought.

Poincaré was able to calculate the orbits of three bodies numerically – we would nowadays do this by computer; he had to do it by laborious calculations on paper – but the orbits were 'so tangled that I cannot even begin to draw them'. Moreover, Poincaré found that when the three bodies were started from slightly different initial positions, the orbits were entirely different. 'It may happen that small differences in the initial positions may lead to enormous differences in the final phenomena. Prediction becomes impossible.'

Poincaré's work has been confirmed by modern mathematical techniques. The description mathematicians would use now is that planetary orbits are 'chaotic'. If you start with the planets in a particular configuration, you can calculate where they will be in, let us say, 100 million years. If you displace one of the planets by just one centimetre from its initial position, you might expect the effect that this would have on the positions of the planets after the same length of time of 100 million years to be about the same size, and completely negligible. But, in fact, the planets could literally be almost anywhere else, within the boundaries of possibility, and the outcome could be entirely different from before. The displacements in position that arise as a result of the slight initial displacement grow uncontrollably.

In modern physics, 'chaos' is the word used to describe behaviour like this, which is predictable in the short term but which, in the long term, depends so much on the initial state that you cannot calculate the long term. Meteorologists can usually predict the weather, more or less accurately, one day or one week ahead. However, since nobody can know about the air disturbances from the flapping wings of every butterfly in Brazil, meteorologists cannot predict when or where a hurricane will strike Florida next year – the small unknowable effects of those flapping wings have completely changed the future. This fact of weather forecasting was discovered in 1963 by Edward Lorenz, a meteorologist at the Massachusetts Institute of Technology. If you change the initial data just a little, the forecast weather patterns can be completely different. Lorenz called the problem 'the Butterfly Effect'; James Yorke coined the name 'chaos'. This concept of meteorological chaos was the same concept earlier discovered as a feature of planetary orbits by Poincaré.

What 'chaos' implies for the solar system is that there have been incalculable upheavals in the positions of the planets over the last 4 billion years since our planetary system was formed. These upheavals were unique events, which have given character to each planet of the solar system. What is even more surprising, and, so far, unexplained, is that, to our knowledge, the solar system, as a whole, seems to be unique.

As I write in 2019, there are about 3,800 planets known in orbit around stars other than the Sun ('extrasolar planets'). Planets appear to be common. On average, there is about one planet per star in our Galaxy – half the stars have no planets, half of them have an average of two. The samples are not complete, because finding planets orbiting around stars that are light years or thousands of light years away is hard and astronomers can only find the easiest cases, but they are good enough to be able, with some thought, to discern some generalities about planets and planetary systems.

It seems that the most common planets in the Galaxy are Earth-like, but twice the size of the Earth – so-called super-earths. Our solar system has four Earth-like planets, the Earth being the largest. There is no super-earth: this might be because we never had one, or because we had one and now it has gone. It is not known what favours the formation of super-earths but our solar system might have missed out on it. Or, alternatively, perhaps our solar system made a super-earth that was somehow flung off into interstellar space? What possible event could have happened in the life of our planetary system; an event that was evidently catastrophic enough to doom a super-earth but let our Earth survive?

Another discrepancy concerns extrasolar planets with a mass close or equivalent to Jupiter's. They are common: and we have a couple in our solar system, Jupiter itself and Saturn. Jupiters are the most frequently discovered of extrasolar planets (but, of course, being the largest and most massive, they are also the easiest to find). The surprising thing about extrasolar jupiters is that they are much nearer to their parent star than our own Jupiter. This warms them up and causes them to evaporate. Jupiters are large because they formed in the far, cold regions of their planetary system: so how did extrasolar jupiters get to the nearer, hotter regions, and, if this is ordinary for many planetary systems, why has this not happened in our solar system?

The bottom line is that our solar system has no parallel among the known planetary systems. Astronomy has no fully accepted explanation for this yet.

Astronomy can, however, explain many of the features of our planets, which can be traced back to particular events. Other secrets remain to be uncovered. In the biography of a historical person there may be gaps. So too with the planets.

Before we can begin looking at their lives, we need to know what planets are. Who are the subjects of this book?

The concept of 'planet' has evolved as our understanding has developed and left us with some confusion. Astronomers themselves have made the confusion more entangled by trying to make everything clear.

Originally, in classical times, the word 'planet' meant a 'wandering star', not a fixed one. Fixed stars were lights in the sky that maintained their positions relative to one another (as far as could be discerned with the equipment available at that time in scientific history); but planets changed their positions relative to the fixed stars. There were seven planets as defined that way: Mercury, Venus, Mars, Jupiter, Saturn, the Sun and the Moon.

Then the perception of the Universe changed when, in 1543, Copernicus realised that our Sun is a star, like the fixed stars, the Moon is a satellite of the Earth, in orbit around it, and our Earth, with Mercury, Venus, Mars, Jupiter and Saturn, is one of six planets, in orbit around the Sun. The orbits of the planets are almost circles lying in the same plane. Further satellites were discovered in orbit around the other planets, and further planets (Uranus and Neptune) were later discovered in more distant orbits around the Sun.

The definition of 'planet' was at that time in history clear, based on the positions and motions of solar system bodies. It began to get muddled when the word took into consideration further issues, ones about the nature of solar system bodies. Comets orbit the Sun, but they are not planets. First of all, they have anomalous orbits. Their orbits are eccentric, not near-circles, and their orbits can be skew, not in the same plane as the rest of the planets. But, most significantly, they have a different appearance, which signifies that they have a different structure. Planets and their larger satellites are almost spherical worlds, either with solid surfaces or enclosed with clouds. They support themselves, and as a result have settled into layered spheres, solid and liquid in the middle, gaseous with an atmosphere on the outside, each layer supporting the lighter layers above. Comets are diffuse (the word 'comet' refers to a hairy appearance), and they have tails: they are structured nothing like planets.

Further discoveries were made in the nineteenth century: small bodies, in orbit around the Sun, with orbits that are mostly near circular and coplanar with the main planets, but crowded together between Mars and Jupiter. These bodies are surprisingly small compared with the main planets. Some of them proved to be near-spherical worlds, but many were irregular in shape. They were at first regarded as 'minor planets', but then they were recognised as belonging to a class of orbiting objects, separate from planets in their nature, and another name became current: 'asteroid'.

Then the classification process for the bodies of the solar system started to go seriously wrong. In 1930, Pluto was discovered. A near-spherical world, somewhat like Mars, it orbits the Sun. It had been discovered as a result of a search for the planet presumed to orbit beyond Neptune, so it had been deemed to be a planet even before it was known to exist. However, its orbit is highly inclined to the orbits of the other planets, and eccentric, so much so that it crosses the orbit of Neptune. Doubts about its status as a planet started to creep in. Then, from 1992 increasing numbers of orbiting bodies were discovered beyond Pluto. They were reminiscent of the asteroids, their shapes a mixture of near-spherical worlds and irregular pieces. They were labelled in an accurate, if unimaginative, way as Trans-Neptunian Objects (TNOs).

These properties were brought together with a growing understanding of the origins of the planets, asteroids and TNOs. The planets are the main result of a process in which large bodies had accumulated from a disc that originally surrounded the Sun as it was forming, the so-called 'solar nebula'. The asteroids, comets and the TNOs were detritus left over from this process, and fragments created since that time by collisions of asteroids. This put Pluto in a new light: as much detritus as planet. It was accurately a Trans-Neptunian Object, but doubtfully a planet. This view led to a redefinition of Pluto, a downgrading, if there could be said to be a hierarchy of status in the bodies of the solar system.

Pluto is indeed a body that orbits the Sun and it has sufficient size that it has settled into a near-spherical shape, supporting itself. However, a third property came into play in the definition of a planet, which ruled Pluto out. This definition was adopted in 2006 by the International Astronomical Union (IAU), representing the global community of astronomers. I was one of the hundreds who raised their hands in the meeting in Prague in the Czech Republic that was called to approve it. It was a controversial decision, which received considerable publicity, because of the perception that it reduced the importance of Pluto. A small army of schoolchildren and other people took exception to this. It is astonishing to me that this mattered to the public at large, but it is at the same time gratifying that an arcane astronomical matter was regarded as so important.

The third property, the one in which Pluto fails to make the grade as a planet, is a characteristic not of its orbit, nor of its structure, but of its previous life. To be a planet, declared the IAU, the body, in addition to the right orbit and the right structure, has to have enough of a size that it cleared out its orbit of other bodies, either gathering them up into itself, or capturing them into orbit as a satellite, or possibly throwing them elsewhere. A planet, said the IAU, has to dominate the orbital zone that it inhabits. Pluto does not: its orbit crosses Neptune's and it ventures among the other TNOs. As a result, Pluto is regarded, not as a planet, but as a 'dwarf planet'. The asteroid Ceres is also regarded as a dwarf planet for similar reasons: it has a similar structure to Pluto and is about the same size, but it orbits among other asteroids that it failed to feed on, so it is not a planet.

In our solar system, the word 'planet', without any qualifying word, is currently reserved by scientists for Mercury, Venus, Earth, Mars, Jupiter, Saturn, Uranus, Neptune. 'Dwarf planet' includes the largest asteroid Ceres, Pluto, and some of the other large Trans-Neptunian

Objects. 'Satellites' orbit their planet. Everything else is defined in the same neutral, unimaginative way as TNOs, as a 'small solar system body'.

There is a summary of the variety of bodies in the solar system and of the vocabulary used in relation to them in an appendix at the end of this book. I am a scientist and I suppose that, strictly, I ought to follow the formalities and restrict this book, with its title *The Secret Lives of Planets*, to the eight planets of our solar system that are recognised by modern science. But when considering what I should write about, I concluded that, if I was too formal, I would be leaving out some of the most significant worlds of the solar system, ones that are the focus of lots of astronomical interest at the start of the twenty-first century. So the book includes the eight main planets, but also two dwarf planets and some of the small solar system bodies, like asteroids and meteoroids, and some satellites. It is my selection of the most significant worlds of the solar system. They are the distinctive characters, the worlds with the most colourful personalities, the ones whose lives are, in my opinion, most worth examining.

CHAPTER 2

Mercury: bashed, bashful and eccentric

> ⊜ Scientific classification: *Terrestrial planet.*
> ⊜ Distance from the Sun: *0.39 times the Earth–Sun distance, 57.9 million km (36 million miles).*
> ⊜ Orbital period: *88 days.*
> ⊜ Diameter: *0.383 times Earth, 4,879 km (3,032 miles).*
> ⊜ Rotation period: *59 days.*
> ⊜ Average surface temperature: *167 °C.*
> ⊜ Secret pride: *'I have the most extreme orbit and most extreme temperature range of any of the planets in the solar system.'*

The battered face of an airless planet reveals the history of its life in the solar system, just as the cauliflower ears and broken nose of a retired boxer speak of his past victories and defeats in the vicious pounding of the boxing ring. Mercury is an airless planet, and its cratered surface speaks of a cosmic bashing known as the Late Heavy Bombardment, about 3.9 billion years ago.

It is not easy to find out about Mercury's early life, because Mercury is the shyest planet. It is hard to inspect closely, even now with the technology of the Space Age. In the days when astronomers had to peer at Mercury only with earth-bound telescopes, it was downright difficult to find out anything about it, all because of its position – it is the innermost planet of the solar system, and from Earth we always look towards the Sun to view it. Hiding bashfully in the skirts of the Sun, Mercury is difficult to see near the bright sunlight, and peeps out only briefly from time to time.

Mercury was the messenger of the gods. The planet and the god have the similar characteristic of travelling quickly back and forth. Because it is the closest planet to the Sun, moved strongly by the Sun's gravity, Mercury is, like its divine namesake, fleet of foot: it is the fastest planet in orbit. It revolves around the Sun in only 88 days – it takes Earth 365 days to do the same. From our position in the solar system, we see Mercury switch alternately from one side of the Sun to the other as it orbits. For about a month the planet is seen as an evening star, visible in the evening twilight, low on the horizon after sunset. It is lost for a month in the Sun's dazzle and then is seen as a morning star, before dawn, for a month. After this, it is lost from view again for about a month behind the Sun, before returning to its starting place. It takes 116 days to complete this cycle. (The cycle of visibility of Mercury from the Earth depends on both the orbit of Mercury and the Earth. This is why the period of visibility for Mercury is different from the period of its orbit.)

Greek astronomers at first thought that the evening and morning appearances of this one planet were of two separate planets and had two names for Mercury: Apollo and Hermes. It was reportedly the mathematician Pythagoras who pointed out, about 500 BCE, that the two were identical. Presumably he noticed that they looked similar and moved at similar speeds, and, crucially, that whenever Apollo was visible, Hermes wasn't, and vice versa. The name Hermes prevailed for the planet; it is still called by that name in modern Greece. In English, the international language of science, the planet's name became Mercury, the Roman equivalent god to Hermes.

There is a connection between the characteristics of those planets known in antiquity, their namesake gods and their supposed effect on people through the influence of astrology. Mercury moves quickly; Venus is the beautiful goddess of love; Mars is a warlike red in colour;

Jupiter, also known as Jove, was the king of the gods, renowned for playing tricks on his subjects; Saturn is slow moving. The English language gets some words for human qualities from these planetary, divine and astrological characteristics – mercurial, venereal, martial, jovial, saturnine. These words are fossil relics of astrology.

At different times in the history of different ancient cultures it was believed that the planets were like the gods, the homes of the gods, the medium by which the gods influenced human affairs, or the gods themselves. We would think that the first two beliefs are poetry. The belief that planets influence the character we have as individuals or what is going to happen to us is astrology, a superstition that is very much alive. And, finally, the belief that planets are deities, or associated with deities, is the religion of astrolatry, which is all but extinct.

If you read a sky guide for amateur astronomers to see when Mercury is visible, you'll probably find a warning note to avoid looking at Mercury through binoculars or telescopes when the Sun is above the horizon. The risk is that you might accidentally pan the short angle from the planet to the Sun while still looking through them. If you look at the Sun, even with your unaided eyes, you can damage them; all the more so if you look through a telescope, which concentrates not only the light but the heat. Professional astronomers can, with planning, take the risk because their telescopes are rigidly controlled. They take care not to jeopardise their eyesight, although they may choose to risk their equipment. If something goes wrong and equipment is damaged, it can be fixed.

Space scientists apply very strict rules about observing Mercury, even though it is only their equipment that is in danger, because the consequences of a manoeuvre gone wrong could be fatal to their mission. It is, at the least, expensive and, at worst, impossible to mend any damage. The Hubble Space Telescope never, under any circumstances, looks at Mercury. This is to avoid even a little bit of the Sun's heat and light from shining down into the telescope. If they do, the structure might distort as parts of it overheat; this could disturb the alignment of the optics. Even a little bit of the Sun's radiation bounced around inside the telescope structure, or reflected and focused by the telescope's glassware onto a delicate component like an electronic detector, would likely be too much.

This all makes Mercury itself hard to look at, so, before the Space Age, a lot of interest centred on Mercury's orbit. Like all the planets, its orbit is basically an ellipse, a squashed circle. You can draw an ellipse by sticking two pins in a piece of paper, and loosely looping a thread around the pins. Put your pencil into the loop of thread and pull it back so that the thread remains taut. What you draw on the paper when you move the pencil around the pins while keeping this tension is an ellipse. Each pin is positioned at what is called a focus of the ellipse. The Sun is at the focus of the orbit of each planet, and the distance from the planet to the Sun changes during the orbit. The amount of change is called the 'eccentricity' of the ellipse, a number that ranges from zero if there is no change – that is, if the orbit is in fact a circle – to nearly 1 if the ellipse is very long and thin. The eccentricity of the Earth's orbit is 0.017: nearly circular. The eccentricity of Mercury's orbit is 0.21, the most eccentric of all the orbits of the planets. As a result, its distance from the Sun varies a lot: from 46 million to 70 million kilometres (29 million to 43 million miles) – that is, from roughly a third to almost half the distance of the Earth from the Sun.

The orbit of Mercury is very eccentric and, because it is close to the Sun, the gravitational pull of the Sun is also very high. Because its orbit is such an extreme case, Mercury is a good choice as a test for theories of gravity. Using such a theory, astronomers can calculate where a planet will be at any particular time, and a successful theory would do this accurately. Newton's theory of gravity passes most tests with flying colours and describes the orbits of the planets very well. But in the case of the planet

Mercury, Newton's theory subtly and mysteriously becomes slightly but significantly inaccurate. The planet's orbit differs from where Newtonian calculations would place it by a small amount each time it revolves around the Sun, which accumulates to a noticeable discrepancy after some decades. The reason for this was a mystery until Albert Einstein created his General Theory of Relativity.

General Relativity amounts to a theory of how gravity works. Einstein put together his theory by using abstract thought alone, without recourse to much that was practical. But of course, he knew that the theory would stand or fall on whether it connected with reality. At first, he did not find anything real that tested the theory. As a result, he held back from telling people about it. The shy planet Mercury, the planet that shows itself briefly and then quickly retreats, as if not wishing to draw attention to itself by saying something that could provoke controversy, ironically gave the reticent Albert Einstein confidence to go public: what he found out about Mercury sorted out the long-standing problem of its orbit and inspired him to go ahead and expose his work to the scrutiny that it has, ever since, triumphantly withstood.

This discrepancy between the observed orbit of Mercury and theoretical calculations based on Newton's theory had confounded astronomers since the nineteenth century. The discrepancy arose as follows.

As it orbits the Sun, Mercury follows an ellipse, as all planets do. However, the ellipse does not stay oriented in the same direction all the time. The long axis of the ellipse slowly rotates around the Sun, at a rate of about 1.5 degrees per century. This rotation of the orbit is called 'precession'.

All planetary orbits precess. Precession is due primarily to the pull of other planets and the fact that the Sun is not completely spherical. The rate of precession can be calculated by Newton's theory of gravity, which gives near enough the right answer for all the planets except Mercury and Venus. Mercury's orbit has the biggest discrepancy, precessing by a rate that is forty-three seconds of arc per century too small (the precession of Venus's orbit is discrepant by 8.3 seconds of arc per century). One second of arc is 1/3,600 of a degree, so the discrepancy is not much, but it was clear and it was a niggling worry.

The French astronomer Urbain Le Verrier had thought that the discrepancy might be caused by the extra pull on Mercury of an undiscovered planet. He had earlier in his career had a great success in accounting for discrepancies in the orbit of Uranus. He supposed that there was an undiscovered planet outside Uranus's orbit that was pulling Uranus off course. This led to the discovery of the planet Neptune (Chapter 15). He tried to pull off the same coup by discovering another new planet, this time inside

Mercury's orbit. It would be even harder to see this planet than it was to see Mercury, and Le Verrier was not put off when he was unsuccessful in finding it.

For some years, astronomers continued to search for the undiscovered planet at times when they thought it might be passing across the face of the Sun, in a 'transit'. When such an event takes place, the transiting planet is silhouetted as a black, circular spot against the bright disc of the Sun. An amateur astronomer, a country doctor, Edmond Lescarbault, living in Orgères-en-Beauce, between Paris and Orléans, claimed in 1859 that he had seen such a spot transiting the Sun over a period of 4.5 hours. Le Verrier travelled to Orgères to interrogate the doctor. He was satisfied that the observation was genuine and he named the planet 'Vulcan', after the god of fire.

The credibility of the doctor's story diminished when he revealed that, although he had made notes about his observation by writing with a pencil on a wooden tablet that he also used to make notes about his patients, he had subsequently shaved off the surface of the tablet with a plane in order to re-use it. Nevertheless, Le Verrier sponsored Lescarbault in his appointment as a *chevalier* of the *Légion d'Honneur*, its medal on a scarlet ribbon being awarded for professional activity with 'eminent merits'. The discovery of a new planet would certainly count as meritorious and the astronomer who discovered it could expect to be seen as eminent.

The fame and prestige that Lescarbault's discovery brought motivated him to devote himself to his passion for astronomy. He abandoned medicine and constructed a house with an observatory with which to pursue his study of astronomy.

However, in subsequent years, other astronomers and Le Verrier himself failed to find evidence that confirmed Lescarbault's claim. There were several attempts to lay out an orbit for the planet and predict when it might cross the disc of the Sun again, but each time, it failed to turn up. So, Vulcan remained controversial and interest in it died away, except for a period of renewed activity among American astronomers when a total solar eclipse was visible in North America on 29 July 1878. This event would reduce the glare of the Sun because the Sun's light would be shaded by the Moon. Could Vulcan be seen during the time of the eclipse, not by the shadow of its transit across the Sun, but in the way that planets are usually seen, by reflected sunlight?

There were two positive reports, but by astronomers whose reputations were questionable; James Watson viewed the eclipse from Rawlins, Wyoming, and Lewis Swift from Denver, Colorado. Each astronomer gave more than one account, and there were discrepancies and inconsistencies between the two of them. Other astronomers had seen nothing like a new planet as they viewed the eclipse, and pooh-poohed their accounts, one of them saying that looking for Le Verrier's mythical birds was a wild goose chase.

Evidently, what Lescarbault had seen was merely a sunspot; such a spot would be stationary on the Sun's face and he must have imagined its motion when he decided that it was a planet. Vulcan disappeared from science back into legend. Lescarbault's medal also disappeared, his appointment to the Legion of Honour rescinded. He must have cut a pathetic figure, old and disgraced, having severed his connections with the community to take up a life alone with his telescope, until his death at eighty years of age in 1894.

With no intra-mercurial planets to disturb Mercury's orbit, the cause of the discrepancy between its calculated and its actual positions remained mysterious until the explanation by Albert Einstein in 1915. In General Relativity, gravity is an effect of curved spacetime. The orbit of a planet is not a static ellipse: it precesses even without other planets to pull it off course. It is a natural outcome of the curvature of spacetime around the Sun.

When Einstein calculated the precession of Mercury, he accounted for the missing 43 arc seconds. The precession of Venus is less, because it is further from the Sun, and the Sun does not cause such a large curvature of spacetime.

After developing General Relativity, Einstein realised that it would get a controversial reception because the theory was so completely new, and contained paradoxical concepts, like the 'curvature of spacetime'. He hesitated to launch it into public scrutiny: at that early stage, it was vulnerable to criticism and subject to doubt, or even ridicule. But after Einstein revealed General Relativity, the fact that it could explain the long-standing mystery of Mercury's orbit by taking account of the additional features in his theory of gravity beyond Newton's provided the support that he needed.

Since 1915, General Relativity has kept faith with astronomers and Mercury has kept faith with General Relativity. It describes the path that Mercury follows better than Isaac Newton's theory of gravity does. Mercury has modestly and patiently confirmed General Relativity, year by year.

A striking feature of the way Mercury behaves was revealed in 1965 by studying the planet by radar, bouncing radio pulses off its surface. The radio frequency of the returned radio pulse from a planet that is rotating is slightly altered. This can provide information about the speed at which the planet rotates and hence its rotation period. This technique discovered that, relative to the stars, Mercury turns exactly three times on its axis for every two orbits around the Sun. However, a 'day' is usually defined as one rotation of a planet relative to the Sun, not relative to the stars – for example, the time from one sunrise to the next. The curious form of the lock between Mercury's rotation and its orbit means that Mercury's 'day' is two of Mercury's 'years' long. It has a 'year' of 88 Earth-days and a 'day' of 176 Earth-days.

This strange relationship between Mercury's 'day' and its 'year' is unique among the planets of the solar system. The large eccentricity adds a further curiosity. It means that the distance of Mercury from the Sun changes by a considerable amount. At a certain time in Mercury's 'year' it is over 20 per cent further from the Sun than normal, so the Sun appears 20 per cent smaller than usual and seems to move 20 per cent slower than normal. Moreover, the planet does actually move about 17 per cent slower than normal in its orbit, exacerbating the effect. Half a Mercury 'year' later, Mercury is closer to the Sun and everything is reversed.

On Earth, the Sun rises and consistently progresses across the sky to the west at more or less constant speed, and it stays more or less the same size. On Mercury, the Sun dramatically and noticeably changes its apparent speed and size as the 'day' and the 'year' progress. The

Sun changes size as seen from the surface of Mercury from a ball that is twice as large as seen from the Earth, to a ball that is three times bigger. After it rises, the Sun moves mostly westwards, but it can stand still and even reverse its motion. From certain positions at certain times of the 'year' it rises but immediately sets, before rising again.

All this makes it problematic to construct a clock and calendar system for use by an inhabitant of Mercury and I have never seen one; but there is no urgent requirement.

The reason for this curious situation is that Mercury's rotation has been slowed by the Sun. The Sun has got a tidal grip on the structure of Mercury so that the planet rotates in synchronism with its orbit around the Sun. This in itself is not an unusual phenomenon, and there are many planet–satellite pairs, and star–star pairs, that are locked up like this.

However, usually the synchronism that locks the spins and the orbits of two close astronomical bodies is such that the tidal forces equalise the spin period and the orbital period. The Earth–Moon system is like this. The time that the Moon takes to go around the Earth is a month, and the time for it to rotate on its axis is the same. Mercury is very unusual in that it rotates three times while going around the Sun twice.

Tidal locking is something that grows over time – Mercury would have been rotating much faster in the past than it is now and has been slowed down by the tidal attraction of the Sun. In trying to explain why Mercury is tidally locked in a different way than usual, astronomers discovered something they had not suspected: the precise way that tidal locking takes place depends on some accidental features of the configuration of the two bodies at their origin. If things had been different early in the life of the solar system, our Moon might not have been locked with one face perpetually towards the Earth and we would have been able to see all over its surface.

Space exploration of Mercury has been very limited, with visits by only two space probes. Mercury is so close to the Sun that a spacecraft risks overheating; it also suffers from storms of solar particles that damage the electronic equipment, both directly by effects of nuclear radiation, and by electrical sparks provoked by a bath of charged particles. The orbit to Mercury is tricky, because, launched from Earth, the probe has to pick up speed to get to and keep up with Mercury, but then slow down in order to go into orbit around the planet. This costs fuel, which must be carried on board, reducing the capacity of the probe to carry equipment to find out things.

What all these difficulties meant was that Mercury kept most of its secrets until the 1970s and even now it is one of the planets about which we know least. The breakthrough happened when an economical way was found to get a space probe to Mercury. The trick was devised by Giuseppe Colombo, known universally by his nickname 'Bepi', an Italian scientist from Padua. He mapped out potential complicated trajectories so that the probe approaches Mercury via Venus and other planets, in just the right way, at just the right time; it is their gravity, not rocket fuel, that helps the probe speed up and slow down to get to the right place.

This 'gravitational sling-shot' technique was used by *Mariner 10*, the first probe to Mercury, which in a looping orbit made three fly-bys in the 1970s (a 'fly-by' is a space mission to another planet that flies close to the planet without entering into repeated orbits or landing). Unfortunately, although the orbit did what it was supposed to and got *Mariner 10* successfully to Mercury three times, Mercury turned each time to greet its visitor with the same face, so the spacecraft mapped only half of Mercury's surface.

The second probe to visit was called *Messenger*, its name a nod to Mercury's place in the pantheon of gods, as well as a contrived acronym for MErcury Surface, Space ENvironment, GEochemistry and Ranging. *Messenger* took six years to travel to Mercury, using gravitational encounters six times before entering into orbit around the planet in 2011. It was able to map almost the entire surface before running out of consumables and being crashed onto the planet in 2015. A third mission, called *BepiColombo* (to commemorate the Italian scientist), was launched in 2018 and, if all goes well, it will explore Mercury for a year or two around 2024–5.

In size, Mercury lies between a satellite and a planet: being only a third of the size of the Earth and a third bigger than our Moon, it is the smallest of the planets, and its gravity is weak. It is a puny planet that lives in the proximity of the Sun, which acts as a big bully. Most obviously, the Sun's heat pours down on the surface and its temperature nearly everywhere is high. As a result, Mercury has lost any atmosphere that it had at the outset, when it formed. But it has gained some atmosphere since then. Its air consists mainly of hydrogen and helium that Mercury has caught from the Sun. There are also less abundant atoms that have been scraped off the surface by the solar wind. The atmosphere is very thin.

Weak as Mercury is, it puts up a defence against buffeting by the solar wind, in the form of a magnetic field – just as the Earth's magnetic field protects our own planet's surface and atmosphere. But Mercury's magnetic field is not strong enough to deflect solar particles when the Sun is particularly active. During the time known as 'sunspot maximum', the Sun has a lot of spots and uses bundles of its own magnetic field to attack Mercury with furious, windmilling arms. At this time, the solar wind blows strongly enough to overcome Mercury's magnetic field and reach its surface.

Because the atmosphere of Mercury is so thin, there is no blanket of air to even out the temperature on Mercury's surface. The surface temperature ranges from as low as −183 °C (−297 °F) at the poles to as high as 427 °C (806 °F) at the equator during the day. In the night-time, heat radiates away quickly from the bare rocks and the temperature of the surface may fall as low as −200 °C (−330 °F). Because the distance of Mercury to the Sun changes so much as it travels around in its eccentric orbit, the amount of heat and light from the Sun that falls on Mercury changes by more than a factor of two, so the temperature at a given time of the day at any one latitude also varies enormously.

Common metals like lead and tin would melt on Mercury's equator, and plastics would melt or decompose. Obviously, this would be fatal for electrical equipment, with plastic-coated components and soldered wires, unless something was done to mitigate against the problem − and, so far, nothing feasible has been found that will do. Even spacecraft that orbit Mercury ('orbiters') have difficulty coping with the Sun's heat, although they can reorient themselves in space to position themselves behind sunshades. Spacecraft that might land on its surface ('landers') or rove about ('rovers') cannot operate there.

Occasionally, Mercury has an atmosphere of steam. This comes from impacting comets. Comets contain a lot of water, which melts and vaporises when one strikes Mercury's surface. The vapour briefly envelopes the entire planet. The floors of some deep craters near the poles never see direct sunlight, so they are exceptions to the statement that the surface of Mercury is hot in the daytime. The crater bottoms never warm above -160 °C (-260 °F). It is so cold there that some comet water condenses to make icy patches, several metres thick in places, and lasts as ice indefinitely. The ice was first detected by the way it reflects radar pulses, and the icy patches were confirmed by the *Messenger* probe in 2008. It was a surprise for astronomers to find that water can survive as ice on the hottest planet, the one closest to the Sun. It was totally out of character.

Mercury's surface is like the Moon's, heavily cratered. If it was possible to view the landscape of Mercury from some well-insulated landing module, either directly as a brave astronaut or with a remote camera, descriptions of the scenery would be similar to those given to us by the Apollo astronauts (Chapter 5). There are craters of all sizes, the largest being Caloris Planitia, which is 1,300 kilometres (800 miles) in diameter, as large as some of the so-called *maria*, or 'seas', on the Moon – the round grey patches that you can see on the Moon with binoculars, and even with your unaided eye. Like the large craters on the Moon, Caloris Planitia is flat bottomed, filled with lava plains, and surrounded by a ring of mountains up to 2,000 m (7,000 feet) tall. It is located on the equator of Mercury, where the Sun shines most strongly, the hottest area of the planet. Caloris Planitia means 'plain of heat'.

Caloris is more than 250 kilometres (150 miles) in diameter and was made by the impact of a large asteroid, about 100 kilometres (60 miles) in diameter. This was perhaps ten times the diameter of the asteroid that caused the extinction of the dinosaurs on Earth. The impact caused seismic waves to travel across Mercury and the 'mercury-quakes' that followed jumbled the rocks on the diametrically opposite area of the planet. The large mountainous and hilly area that was thus created is called 'weird terrain'. The seismic waves reflected back from the far side of Mercury, filling the planet with seismic sound, so that for hours or days it rang like a bell. The pulse of energy cracked the surface of Mercury, liberating molten lava from its interior. The impact triggered volcanic activity that flooded large areas, which, by contrast with hilly weird terrain, constitute smooth lava plains. The impact shook Mercury's mountains. Landslides slithered down their sides. The planet-wide consequences of the Caloris Planitia strike speak of a terrible impact with global effect. The impacting asteroid shook Mercury to its core. Had it been much larger, it could have been literally world-shattering.

It seems that there were two periods when asteroids rained down on Mercury in abundance. The first was the time of hurly-burly in the aftermath of the formation of the solar system, when the planets were being built up. Planetesimals, or potential planets, attracted and accumulated small solid pieces that had formed in the discarded material of the proto-Sun – material that spun off from the Sun as it condensed – so the solar system then was full of bits of all sizes.

Some of the pieces had amalgamated up to the size of asteroids. But there were a lot of small bits left over, pieces the size of small pebbles or boulders. Some of these bits still survive as rocks orbiting the solar system. From time to time some of these primitive rocks fall on the Earth, meteors of a particular sort called chondrites. They are 4.568 billion years old, as measured by looking at the products of radioactive decay in the rocks. Radioactive elements and their products are trapped in the rocks when they solidify, and decay at a rate that is accurately measured in a laboratory. Astronomers assume that the moment that the rocks solidified was the birth of the solar system. Given how long ago this event was, it is remarkable how precisely its date is known.

During the first cratering period on Mercury, the impacts of these pieces of rock and asteroids created craters of all sizes, from smallest to largest. By contrast, in the second cratering period there were disproportionate numbers of large asteroids, the smaller ones having been used up or consolidated into bigger ones, so the craters were bigger on average. This second period is known as the Late Heavy Bombardment.

The Late Heavy Bombardment occurred about 3.9 billion years ago, about 600 million years after the solar system formed. This age is inferred, not from Mercury but from our Moon. It shows up in the age of rocks collected from the Moon. The rocks come from three sources.

About 300 grams of lunar soil was returned from the Moon in the 1970s by three robotic probes from the Soviet *Luna* programme. Small probes were sent to the Moon and settled base-down on the lunar surface. They each reached out a robotic arm and scooped soil into a small rocket, firing it back to Earth to parachute onto the Russian steppes.

Around the same time, the *Apollo* astronauts collected about half a ton of lunar rocks into numbered bags using tongs and scoops, packed them into suitcase-like aluminium containers, and personally escorted them back to the USA.

Other lunar rocks have been found among meteorites. Three hundred pieces of the Moon have been found that fell to Earth after being knocked off the Moon's surface by the impact of asteroids.

The oldest lunar rocks are those collected from the lunar highlands, the lighter areas of the Moon. Individual rocks from the lowlands, the dark *maria*, have ages that seem to cluster between 4.0 and 3.85 billion years. This was when they last solidified. It appears therefore that the crust of the Moon was strongly heated 3.9 billion years ago. This was discovered between 1974 and 1976 by a group of Sheffield University astronomers led by Grenville Turner. They suggested that, after the Moon had first solidified about 4.5 billion years ago, asteroids heavily bombarded the surface of the Moon for 200 million years starting 4.0 billion years ago and re-melted it. The Sheffield group called this event the 'Lunar Cataclysm' – an early name for the Late Heavy Bombardment.

The reason that the bombardment occurred is unsolved. There may have been a major collision between two large asteroids or planets, causing a number of fragments, including some very large ones, to spray through the solar system, impacting everything they encountered. Another possibility is that the asteroids, which were, up to that time, generally orbiting peacefully, were disturbed by the movement of giant planets, Jupiter and Saturn, and scattered everywhere. According to one theory known as Grand Tack, when the giant planets were still orbiting in the disc of debris left over from the formation of the solar system, Jupiter moved further in towards the Sun under the influence of the debris. If uninterrupted, this migration would have left Jupiter in an orbit much closer to the Sun, which would have left our solar system like many recently discovered planetary systems. Many exoplanetary systems have so-called 'hot jupiters' – gas giant planets that must have formed far out in their planetary system but have migrated inwards. They are now much hotter than they were, with their gaseous material evaporating and dissipating. Our Jupiter avoided this fate because, in some way (see Chapter 9), it reversed its course and tacked like a boat to sail against the tide, returning to a final orbit further out. Along the way, it scattered debris and asteroids, sending bits of rock flying towards Mercury, but also the Earth and Moon.

A third and, in many ways, the most interesting scenario for the Late Heavy Bombardment has been put forward as a result of what is known by astronomers as the 'Nice

Simulation'. This theory has become popular because it offers the prospect of explaining several secrets in the lives of the planets – one theory, several explanations: that is powerful as well as economical!

Nice is pronounced 'niece', because the work to create the simulation took place in 2005 in the French city of Nice by an international group of mathematicians centred on the Côte d'Azur Observatory and led by Alessandro Morbidelli. According to the Nice Simulation, what happened in the first billion years or so of the history of the solar system was like a gigantic game of interplanetary billiards or pool played by hyperactive children let loose around a billiard or pool table.

The Nice Simulation is one of a number of calculations of how the planets might have interacted, at the stage at which the planets had just formed in the solar system. Because of the limitations imposed by 'chaos theory' (see Chapter 1) it is not possible to know exactly what happened that far back in the past. So, it is impossible to know where exactly the planets lived their lives in the distant past. That is a secret from their youth which they will retain.

What can be done is to make simulations: these are calculations about a large number of possible scenarios, of various degrees of invention, from small changes in the details to wholesale changes in the architecture of the solar system, such as the number of planets. Astronomers can then see which simulations fit best to what they know. The more plausible features of the simulations are the ones that recur in a large proportion of the calculations. These are taken to be something close to what actually happened. The distillation of all these attempts to calculate what went on is known as the Nice Simulation.

The Nice Simulation starts at a time when almost all the stuff from the interstellar cloud that formed the Sun had been blown out of the solar system except for bits of solid material. The solid lumps were in orbit around the Sun, much like the planets, comets and asteroids now, but there were more of them, and they were everywhere. The bits of solid material that lump together to form planets are called planetesimals and a lot of planetesimals had been created in this process. They moved among the planets. The planets at that time included the four outer, giant planets that we know today (Jupiter, Saturn, Uranus and Neptune) but perhaps more than half a dozen inner 'terrestrial planets'. More than half a dozen is a few more than the four we now have (Mercury, Venus, Earth and Mars). The giant planets were near to their current orbits; but perhaps there were more of them as well, five or even six, not just four.

According to the Nice Simulation, there were occasional close encounters between the planetesimals, and between individual planetesimals and the larger planets. Some of the planetesimals were ejected from the solar system, perhaps the vast majority – these now constitute interstellar asteroids, little worlds travelling for ever in the cold darkness of space, far from the light and warmth of the Sun, orphans lost in the empty spaces of the Galaxy.

The same thing may well have happened to other planetary systems. Perhaps some time in the future, one of their planets will loom up out of interstellar space and speed through our solar system. This may have already happened. There are a few asteroids that orbit backwards, and some astronomers speculate that they may have been captured from space. Then, in 2017, the Pan-STARRS telescope on Hawaii discovered an asteroid that really was caught in the act of falling at unusually high speed into the solar system from outside.

One hypothesis was that this errant body was a comet, but it showed no signs of developing a coma or a tail. It proved to be unusually long and thin, changing brightness as it rotated, dimmer when presenting a small area when it was end on, brighter when seen from the side. A second hypothesis was that it was an interstellar asteroid, attracted into the solar system by the pull of the Sun.

A third idea was that it was long and thin because it was an interstellar spaceship. Although this idea seems fanciful, it was given additional credence by the discovery that its orbit was not controlled solely by gravity – there was an extra force on it, from a sort of propulsion system. Some astronomers argue that the body really is some kind of a comet, its high speed due not to its interstellar origin but to the rocket-like effect of backward-facing fountains of gas that spray out and boost it along. One astronomer has suggested that it has a so-called 'solar sail'. This is a device that has been proposed by terrestrial space engineers to take advantage of the push that comes from sun- or star-light shining onto a large reflecting sail. Perhaps extraterrestrial beings had constructed such a device to power a spaceship to visit and explore other planetary systems in the Galaxy, including ours.

Whatever the truth of this far-fetched suggestion, the visitor is now speeding back out into space, and will not be back. Some similar visitors have already come in and settled in the Sun's domain, masquerading as asteroids. This visitor would be the first seen to sweep past like a sailing boat in windy conditions failing to make it into harbour. The name that the body was given reflected the conviction that it was of interstellar origin. The Pan-STARRS telescope is in Hawaii, and the astronomers there consulted the local community for suggestions. The body was named "Oumuamua', which, in Hawaiian, means 'the first messenger to arrive from afar'.

'Oumuamua made its closest approach to the Sun and then zoomed by the Earth, within 24 million kilometres (15 million miles). This is about sixty times the distance from our planet to the Moon – no distance at all on a cosmic scale. If any future interstellar visitor turns out to be of significant size, it could rampage through the solar system, disturbing the orbits of the planets in a minor way, or a major one, depending on how close its trajectory takes it, with unpredictable consequences.

When planetesimals were being ejected from our solar system, they gave the planets remaining in the solar system a little backward kick. The giant planets gradually migrated in towards the Sun. After tens or hundreds of millions of years this brought the two innermost giant planets, Jupiter and Saturn, into resonance, with two of Jupiter's orbits taking exactly the same time as one of Saturn's. This Jupiter/Saturn resonance is called a 2:1 resonance, pronounced 'two to one'. It put the two planets into a condition that had a profound effect on the other planets and the myriad smaller bodies of the solar system, the bits and pieces left over from the process of planet-building. The effect comes from the nature of a 'resonance'.

When a parent pushes a child on a swing, this is an example of a resonance. The child starts swinging, is pushed by the parent, and swings away further. On the child's return, the parent pushes away again, and so on. The amplitude of the swing builds up, as the child wanted. The push does not have to happen every swing. The same effect would happen if the parent pushes the child on alternate swings in a 2:1 resonance. The essential thing is that each little push comes at the same point in the cycle of the swing. When two planets resonate, they likewise create a gravitational force field that has the same effect over and over again, and this can build up a major disturbance on a third planet nearby.

At the outset of the simulation, Jupiter and Saturn were almost in resonance. They were brought nearer to resonance and then into resonance by the random ejection of some of the planetesimals. The augmented gravitational force field that they then created affected all the other planets. Some more of them were ejected into space. The outcome for the terrestrial planets (the rocky ones near to the Sun) was that just four were left behind, the four we know today (Mercury, Venus, Earth, Mars).

There was at that time a counter-factual future life for the Earth, in which the Earth became an interstellar planet, roving around the Galaxy like a lone coyote on the icy, vacant prairie. This did not happen to our own planet but it may well have happened to one of Earth's former neighbours. Perhaps this is what happened to our solar system's super-earth, if we had one.

Whatever might have happened to eject our Earth from the solar system, it didn't. However, in this chaotic period in the development of the solar system the Earth shifted its orbit back and forth towards and away from the Sun. Our planet ended up in the Goldilocks Zone of the solar system, which made the evolution of life possible. It was a matter of luck.

The rest of the solar system was also affected. The asteroids were pulled and swung out of their orbits, heaved about by the massive effects of Jupiter and Saturn. Some asteroids, jaywalking across the more orderly circular paths of the planets, fell on the planets, especially those inwards towards the Sun, like Mercury. They pounded their surfaces, making craters – perhaps this was the event that we know as the Late Heavy Bombardment.

Of course, if Mercury suffered under the Late Heavy Bombardment, so did the Earth and Moon. The event produced about 1,700 craters on the Moon larger than 20 kilometres (12 miles) in diameter, and, mathematically, there would have been ten times this number produced on Earth – some would have been 1,000 kilometres (600 miles) across. The terrestrial craters are now eroded away by 3.9 billion years of weather, but there is some evidence that the Late Heavy Bombardment happened on Earth in the composition of deep ocean sediments. Sedimentary layers laid down at this time in Greenland and in Canada survive and can be analysed.

There are differences in composition between material of extraterrestrial origin and material originating on the Earth. Some chemical elements are more abundant in meteoritic material than that from the crust of the Earth. Another difference is in the isotopic composition: isotopes are varieties in the nuclear composition of chemical elements, and the proportions of different isotopes reflect the chemical processes that produced the material. The composition of the Greenland and Canadian sediments from 3.9 billion years ago suggests that they contain more meteoritic material than usual, brought to Earth in the Late Heavy Bombardment.

It might also be significant that the fossil record of life on Earth seems to have started after 3.9 billion years ago – if life had evolved before this it may have been set back badly by the Late Heavy Bombardment and earlier traces erased. Alternatively, it may have been the Late Heavy Bombardment that set off the evolution of life by bringing an abundance of water and organic molecules on asteroids to the Earth's surface, the water warmed in the bombardment. This would have been the time in the history of the Earth that Charles Darwin was referring to when he wrote (in 1871) that the origin of life could have been 'in some warm little pond ... that a protein compound was chemically formed ready to undergo still more complex changes . . .' The 'little pond' was the ocean on the primitive Earth, its water and organic chemicals brought to Earth in asteroids and comets and warmed by the energy of the bombardment and by geothermal energy.

It turns out, then, that the secret life of Mercury, written on its crater-scarred face, is a clue to the secret life of our own Earth and to the secret of life itself.

CHAPTER 3

Venus: an ugly face behind a pretty veil

- Scientific classification: *Terrestrial planet.*
- Distance from the Sun: *0.72 times the Earth–Sun distance, 108.2 million km (67.2 million miles).*
- Orbital period: *225 days.*
- Diameter: *0.949 times Earth, 12,104 km (7,521 miles).*
- Rotation period: *243 days.*
- Average surface temperature: *464 °C.*
- Secret wish: *'I hope that the climate change under way on my twin is not as extreme as it was for me.'*

Venus the goddess was the epitome of beauty; there is likewise little in the celestial sky to rival the glorious sight of Venus the planet. Shining white as the Evening Star against the yellow, orange and red of a sunset, below the deep violet of fading daylight, Venus is unmistakeable, both in its pure colour and in its radiant brightness.

Like Mercury, Venus goes around the Sun in an orbit that lies between us and the Sun, so it switches from side to side of the Sun, completing one cycle of appearance in 584 days. Thus, Venus is an evening star for several weeks at intervals of about one and a half years. In ancient times, like Mercury, Venus had two names, as if it were two separate planets. It was Hesperus or Vesper (in the evening) and Phosphorus or Lucifer (in the morning). Then the penny dropped in the Hellenic scientific world with the realisation that these were apparitions of a single planet.

The morning apparitions of Venus are as beautiful as the evening ones; in fact, the clarity and stillness of the air in that magical, chilly half-hour as dawn breaks add a purity to the morning appearances of Venus that makes them even more special. If I see Venus in the dawn as I leave my telescope after a night's observing at an observatory on a cold mountaintop, my heart lifts and I forget how tired I am.

Venus is the brightest celestial object in the sky after the Sun and the Moon, and, occasionally, an exploding star (supernova). Apart from these three cases, only the reflections from the *International Space Station* (*ISS*) rival Venus; it somewhat saddens me that, at its brightest, the man-made *ISS* can, for the brief period of its passage across the sky, somewhat surpass and distract from the natural glory of the planet.

Venus is thus the brightest planet. Three of the four reasons for this are that it is quite large, it is near the Sun and it is near us: it intercepts a lot of light from the Sun and the light that is reflected is not diminished much by its travel over the short distance to us. The fourth reason is that it reflects a large fraction of the sunlight that falls on it. That is because Venus is completely enveloped in white cloud. The cloud reflects three-quarters of the sunlight that falls on it. This is several times the average reflectivity of the other planets. But the cloud cover means that the beauty of Venus is the beauty of a veil that completely hides the ugly reality of her scarred face.

I can remember as an amateur astronomer in my school days scrutinising Venus night after night with my little, home-made telescope to discern any of the planet's secrets. It looked most like a white billiard ball, illuminated from the side. I strained to see through the cloud to the imagined surface below. Some people can from time to time see with their telescopes weak shadings over the surface of Venus, but the only features that I could ever see were slight irregularities in the boundary between the illuminated half and the dark half of the planet. These, I learnt, were shadows cast by the tops of clouds of different heights. I thrilled to know this small fact about this veiled world. It was not as dramatic as glimpsing the surface but I imagined flying through those towering clouds, the stars above me and Venus hidden below. It might have been thoughts like these that inspired me to become an astronomer, even though I suffered the scorn of my school friends, who wondered what I found so compulsive about the sight of a white sphere.

My little telescope at that time had the advantage of a modern lens but was only a bit better than the one that was used by the Italian physicist Galileo Galilei to view Venus in 1610. I saw what he saw. Galileo's telescopes were primitive by any modern standard, with small, unsophisticated lenses and supported on rickety table-top stands. Nevertheless, the light-gathering power and the magnification of these telescopes were a large improvement on the performance of the human eye. Two of his telescopes survive in a museum in Florence, the lens of one of them, now preserved in an ornate ivory display case, cracked in a moment of careless handling by the staff of one of the Medici family. The lenses have an aperture of 40–50 mm, compared to the typical 6 mm for a dark-adapted eye, and they magnified fifteen to twenty times. Galileo could see considerably more than anyone had seen before.

Galileo observed that at its greatest distances to the side of the Sun, Venus showed a half-moon shape with the bright side facing in the Sun's direction. As Venus moved towards the Sun, however, its shape either moved towards full circular illumination or narrowed to a thin crescent depending on whether it was passing behind the Sun or in front. This discovery would place Venus firmly at an important point in the history of science.

Galileo published what he saw in a Latin anagram, which he sent from Padua to his fellow astronomer Johannes Kepler in Prague. The anagram spread throughout the scientific world. The coded sentence was *Haec immatura a me jam frustra leguntur o.y.* It can be translated as 'Things not ripe for disclosure are read by me'. (The letters o. and y. at the end are odd letters that Galileo could not make work in the anagram.) The anagram's letters can be rearranged to *Cynthiae figurae aemulatur mater amorum.* This means 'The mother of lovers imitates the shapes of Cynthia', or in plainer English, 'Venus has phases like the Moon'.

The anagram method of making a coded announcement was common in the seventeenth century, to help establish priority for a discovery. The anagram was promulgated from person to person at the slow pace at which, in that time, news spread. When the solution was announced by its author, everyone was then able to recognise who had made the discovery.

Simultaneously with his observation of the phases of Venus, Galileo discovered something that had been mentioned by Copernicus in 1543, in his book about his model of the solar system in which he said that the planets moved around the Sun – the heliocentric theory of the solar system. He pointed out that the size of Venus should appear to change as the planet came closer to Earth and got further away. At its furthest, beyond the Sun, it would be well over four times further than at its closest, when it lay between the Earth and Sun. Galileo discovered that Venus did indeed change size according to its position – it was smallest when it was showing a full face, and it was biggest when it was crescent-shaped.

The significance of Galileo's discovery of the phases of Venus was that it showed immediately that the planet went around the Sun, in an orbit that lay inside the orbit of the Earth. When Venus was on the far side of the Sun, it was facing directly into the Sun and to the Earth. Its face was fully illuminated like the Full Moon, and it was small because at its most distant. By contrast, when Venus was moving between the Earth and Sun, its face was towards the Sun and its un-illuminated rear was towards the Earth, so it showed as a large, thin crescent. This was exactly what Copernicus had suggested. Galileo had proved Copernicus's model of the solar system. What he had seen could not fit with the earlier model of the solar system, the geocentric model. This claimed that Venus orbited the Earth in a circular orbit below the Sun. If that were the case, Venus would be the same size all the time.

Galileo knew that his discovery would be rubbished by some people. On New Year's Day 1611 he published the solution to the Venus anagram, explaining that the phases meant that it must orbit around the Sun:

> ... something indeed believed by the Pythagoreans, Copernicus, Kepler and myself, but not proved as it is now. Hence Kepler and other Copernicans may glory in their successful theories, although as a result we will be thought to be fools by most bookish philosophers, who will regard us as men of little understanding or common sense.

Unfortunately, Galileo correctly foresaw the controversy that his thoughts would generate. He was summoned by the Inquisition, tried and labelled as a heretic for teaching things that were contrary to Scripture. Made to recant and banned from doing anything to communicate his astronomical discoveries, he was placed under house arrest until his death in 1642. However, he was right, and Venus has retained its critical place in scientific history.

The phases of Venus are not intrinsic to the planet: they are consequences of the orbit of Venus, and its position relative to the Sun and us. What is the nature of Venus itself? Because of the unbroken, thick cloud, little was known before the twentieth century. This gave free rein for astronomers to imagine what its surface might be like under the clouds. Venus is about the same size as the Earth, and was regarded as the Earth's Twin. It evidently has an atmosphere. The planet lies at a distance from the Sun that is 75 per cent of the Earth's distance, closer to the Sun than Earth is, so it must be warmer, and, on the assumption that the clouds were made up of water droplets like terrestrial clouds, the climate was thought to be humid. Thus, to many astronomers in history Venus was the most likely of the other planets of the solar system to harbour life. This all suggested that the climate had similarities to that in the hot countries of Earth, with inhabitants likewise similar. A book of 1686 called *Conversations on the question whether there are other worlds* by the French writer Bernard de Fontenelle, translated in 1700 by Aphra Behn, one of the first English women authors and playwrights, brought the patronising, racist stereotypes of the time to thoughts on the potential inhabitants:

[T]he climate, being nearer the sun than we, receives from its influence a brighter light and a more enlivening heat ... [T]he inhabitants of Venus are ... little sunburnt gentlemen, always in love, full of life and fire, given to making verses, and great lovers of music, and every day inventing feasts, balls and masquerades, to entertain their mistresses.

This picture of an Earth-like Venus proved to be far from the truth. In reality, there is a dark secret below those pure, white cloud tops. They hide an ugly face and a hellish disposition in which it is hard, even impossible, to imagine anything living. The surface of Venus is covered by sterile, scaly, black volcanic rock, cut by frozen rivers of lava that have flowed from numerous volcanoes. Its atmosphere is very dense – the atmospheric pressure on the surface is ninety times that on Earth, equivalent to the pressure at a depth of about a kilometre into the ocean. This is about the same as the maximum depth at which submarines usually operate on Earth, although specially constructed research submarines and rescue vessels can go deeper. The composition of the atmosphere is mostly carbon dioxide, with nitrogen and traces of sulphuric acid, hydrogen chloride and hydrogen fluoride. These acids drop to the surface in acid rain that ensures a quick death for machinery like spacecraft that land on its surface. The clouds are floating droplets, not of water as on Earth, but of sulphuric acid. Dim sunlight filters through the thick, sulphurous atmosphere to cast an evil, yellow colour onto the surface from above.

Few details of this picture of the surface of Venus were known before the Space Age; just the overall composition of the atmosphere, its high density – and its temperature, which is literally the killer fact, so far as life on Venus is concerned. In 1956, pioneer radio astronomer Cornell H. Mayer and his colleagues at the US Naval Research Laboratory collected microwave observations of Venus – these longer wavelength radiations come from deep within the atmosphere. They discovered indications that the surface has a very high temperature, much higher than Earth, hot enough to melt lead.

The first space visitor to Venus was the US *Mariner 2*, a fly-by in 1962. The new rocket with which it had been intended to launch *Mariner 2* ran into problems during its development and a standby of a previous type had to be used. It was smaller and its carrying capacity was less. As a result, the equipment on *Mariner 2* was greatly reduced from the original plan. Nevertheless, the mission was a success. The radio investigations of six years earlier had probed through Venus's atmosphere from afar and had not been fully definitive. *Mariner 2*'s major scientific result was that it confirmed from close up that the temperature of the surface of Venus is very high, over 400 °C (750 °F).

Further *Mariner* probes followed in the 1960s and 70s. But the most detailed knowledge of the nature of the surface of Venus came from a series of Soviet landers, parachuted below the clouds. NASA in the USA in the 1960s had concentrated on the *Apollo* programme to land human beings on the Moon, in accordance with the 'We choose to go to the Moon' challenge issued by President Kennedy in 1962. Meanwhile, the Soviet Union made the closest planets, Venus and Mars, its primary targets for space investigation. In 1967 the Soviet *Venera* spacecraft series, initially designed by pioneering Soviet space engineer Sergei Korolyov, began their scientific exploration of Venus. The early *Venera* spacecraft were, like *Mariner 2*, remote-sensing, fly-by missions. These were followed by a number of spacecraft that entered the atmosphere of Venus on parachutes to measure its properties directly for about an hour as the probes descended. The probes were buffeted by strong winds, at speeds of up to 200 mph. In fact, the atmosphere of Venus shows superfast rotation, by contrast to the solid planet. The planet rotates once every 243 days, backwards. The upper regions of the atmosphere rotate once every four days at the equator.

The only spacecraft that have landed on the surface of Venus were *Venera* probes, made for the Soviet space programme. The probes were made in Moscow by a space science institute called the Lavochkin Association, associated with a Russian aerospace company with a similar name that makes warplanes. On a visit, I was shown its private museum, which houses equipment from the Soviet-era space programme. I looked in awe at a cannon-ball-like capsule from the *Luna* programme. It was the very one that had been taken to the Moon on a lander, which packed it with lunar soil and then blasted it off to return to Russia. It was blackened and dented from its fiery, high-speed descent through the atmosphere onto the Russian steppes. But it was strong and stayed shut so as to offer its small cargo, uncontaminated, to Soviet moon-scientists for analysis. The *Venera* landers (what I saw were spares – the flight models are still on Venus) were very robust-looking, strongly riveted together. They reminded me of railway machinery rather than space-craft. They had been made by engineers who fully understood the harsh conditions into which they would be parachuted, including a crushingly high pressure.

The first probe to land on the surface of any other planet was *Venera 7* in 1970. Something went wrong during the descent to Venus, which was much faster than planned. It struck the surface at the speed of a car in a serious road crash, damaging equipment. But because the lander was so strong, it was not destroyed. It fell over, ending up on its side with its radio antennae pointing away from Earth. Nevertheless, a weak radio signal provided twenty minutes of information about the surface conditions on Venus. *Venera 8* was the first in the series (1972–84) that landed successfully on the surface. Typically, each landing mission survived on the surface for up to an hour before the severe heat and atmospheric conditions caused the equipment to fail.

These missions showed that, although Venus's thick clouds do not extend down completely to the surface, the view over the landscape is limited. The landers could see hills only as distant as 3–5 kilometres (2–3 miles). At the landers' feet, they saw crazed and scattered plates of black, volcanic rock. The nature of the wider landscape remained a mystery.

Exploring Venus by viewing it in the usual way with ambient light has strong limitations. But radar can penetrate clouds, and this set the method for future exploration of the surface of Venus. After some early trial investigations, there was considerable progress with two *Venera* missions (*Venera 15* and *16*, both in 1983), which were equipped with radar systems that allowed good coverage, with a spatial resolution of about 1 kilometre (about 1,000 yards) and a height resolution of about 50 metres (about 150 feet). These missions saw recognisable volcanic features on Venus, as well as some new kinds. There are shield volcanoes and numerous volcanic cones, structures that are well known from examples on Earth.

There are also structures not found on Earth, such as 'coronae', large circular structures that, up until then, had been mistakenly identified as lava-filled meteor craters. There are 'arachnoids', spider-like collapsed domes with radial cracks that look like legs. Both these structures mark stationary hotspots under the surface that generate volcanic blisters in the planet's crust. Venus does not have moving tectonic plates that collide, and so the volcanoes of Venus do not line up in rows as they do on Earth in the Ring of Fire around the Pacific Ocean; they are dotted about almost everywhere.

The breakthrough in knowledge of the surface of Venus came with a remarkable mission, NASA's first interplanetary mission for eleven years, launched from the Space Shuttle in 1989, that reached Venus in 1990 and remained in orbit around the planet for four years. It was called *Magellan*, after the explorer who mapped the Earth. The body of the spacecraft (the 'bus') was ingeniously assembled at low cost from spare parts left over from previous missions. The bus was equipped with radar to map Venus. Challenged by the *Venera* missions, NASA upgraded the planned resolution from 0.6 kilometres (600 yards) to 100 metres (100 yards), so *Magellan* could see geographical details as small as the size of a soccer pitch. The spacecraft orbited around Venus, not around its equator but over its north pole, down and under the south pole, repeatedly every three hours. The plane of the orbit stayed fixed in space while the planet rotated underneath. The spacecraft's track gradually covered almost the entirety of the planet. It mapped the successive strips, creating a radar image that showed the height and roughness of the terrain – its 'geomorphology'. Venus stood unveiled in all her glory, its secrets on plain view.

Magellan revealed that Venus is almost entirely – three-quarters – covered in a volcanic landscape. There are about a hundred shield volcanoes similar to Mauna Loa on Hawaii and hundreds of thousands of smaller, isolated volcanic cones like Mount Etna in Italy. The highest volcano is called Maxwell Montes and reaches 11 kilometres (36,000 feet), comparable to Mauna Loa as measured from the sea floor. Generally speaking, the volcanoes on Venus are larger in area, but no higher, than the volcanoes on Earth. Rivers of frozen lava have flowed down valleys on the volcanic slopes, and show up as sinuous lines of a rough jumble of craggy rocks, like recent lava flows on Earth. But the lava flows can be very much longer and wider than terrestrial ones. These slight differences are probably down to the surface rocks having a different composition, with the rocks on Venus being somewhat more malleable.

About 10 per cent of Venus is loose, dusty material that has accumulated in the lowlands, covering up what is underneath. It has been generated by erosion of the highland areas, caused by chemical reactions between the rocks and the atmosphere. There is not much water on Venus, just a little in the atmosphere, so there is no erosion by water to the same effect as on Earth. There has been some erosion by the wind blowing dust so as to abrade the remaining rocks, and by meteor impacts, which break up the rocks of the surface and spray rubble everywhere.

Meteor craters are as rare on Venus as volcanic craters are abundant – only about a thousand have been identified, many fewer than on airless Mercury and the Moon. This is, however, many times the number known on Earth. Terrestrial craters are destroyed by the effects of erosion by the weather and the way that rocks are churned over by the shift of tectonic plates. Venus has weather but not tectonic plates. In part, the small number of craters on Venus is because the atmosphere of Venus defends the surface against meteor strikes – meteors tend to burn up as they fly through the atmosphere. Indeed, a number of meteor craters form groups, as if they were made by a meteor that broke into pieces but was not entirely destroyed, so that a number of the pieces impacted together as part of the same strike.

But the main reason why there are so few craters is that, as on Earth, the surface of Venus is young (as planets go), even though there are no tectonic plates sliding about as on Earth. There may have been more craters in the past on Venus but they have been overlaid with fresh volcanic material, and lie beneath the surface. Astronomers can relate the number of craters to the rate at which asteroids strike the surface of Venus at the present time, and estimate that the surface is perhaps half a billion years old. Three or four billion years or more of the history of Venus lies hidden, a secret under the new surface, a past erased like that of a defecting double agent given a safe house.

Has the resurfacing happened throughout the last half-billion years? Or was the whole planet resurfaced in one big event half a billion years ago? There are likely some volcanoes active on Venus now, with some recent ash flows spotted on one or two volcanoes. Some 'hot spots' on the side of some volcanoes were identified by a recent European space mission called *Venus Express*. It looks as if these are heated by liquid lava in lava pools below the surface, but no actual eruptions have ever been seen. Venus is geologically active, but not very. This suggests that the period of intense volcanic activity was a long time ago.

The volcanic activity which resurfaced Venus was a catastrophic event. What provoked it? Was the entire planet consumed by volcanic fire, in an expression of internal rage? Or was the volcanic inundation provoked by something external? So far, Venus has kept this entire phase of its life a mystery.

The volcanic resurfacing of Venus was not the only global catastrophe suffered by Venus in its life. Although there is now no sign of water on the planet, it seems likely that, when it was formed 4.6 billion years ago, and as earlier astronomers had speculated was possible, Venus was not very different from the Earth. Probably all the terrestrial planets (Mercury, Venus, Earth, Mars) were born in a similar state: all new-born babies look alike! Like planets, they grow to become distinct individuals when small differences in their nature are amplified as they age, and as they are nurtured in different circumstances.

If it is reasonable to assume that Venus had liquid water on its surface, brought there by asteroids and comets, it probably also had an atmosphere of nitrogen, water vapour, carbon dioxide and methane from volcanic emissions. The carbon dioxide and the methane created a strong greenhouse effect, which was augmented by water vapour as the oceans evaporated because of the proximity of the Sun.

The greenhouse effect is a property of some of the gases in planetary atmospheres. It is important for us on Earth. It keeps the surface of the Earth at a comfortable, relatively constant, life-sustaining temperature.

The greenhouse effect in the Earth's atmosphere was

discovered in 1827 by Joseph Fourier, the French mathematician and physicist. The Earth receives light from the Sun, of which about 70 per cent is absorbed. This warms the land, atmosphere and oceans. These warm masses radiate heat back towards space in the form of infrared emissions. But the infrared emitted from the surface is mostly absorbed in the atmosphere by greenhouse gases and clouds. It does not escape into space, but heats the lower air. This impedes the outward flow of heat from ground and ocean. The analogy is that, as in a greenhouse, the surface and lower air layers of the Earth get hotter as a result.

From the outset, the greenhouse effect on Venus was stronger than on Earth. It warmed the planet so much that its oceans completely evaporated. This occurred long before the volcanic resurfacing of Venus, so all traces of the oceans, such as the flow of water in streams or floods, or rocks laid down in minerals formed in water, have been covered over. The extra water vapour in the atmosphere enhanced the original greenhouse effect and increased the temperature of Venus further. This changed the composition of the atmosphere, even making more greenhouse gases, like clouds of sulphuric acid droplets. This raised the temperature still further – and so on. Cornell University astronomer Carl Sagan put forward in 1961 this explanation for the very high temperature of Venus – the greenhouse effect on Venus ran away.

The greenhouse effect completely dominates the climate of Venus, raising its temperature by 500 °C (900 °F). By contrast, the greenhouse effect raises the temperature of the Earth by just 33 °C, very significant for the Earth's climate, but beneficial and not catastrophic. However, man-made ('anthropogenic') greenhouse gases liberated by industrial and agricultural activity (carbon dioxide from burning fossil fuels and methane emitted by cattle) are increasing the temperature and threaten to upset this benign equilibrium. The Paris Accord of 2015 sets out the steps necessary to keep the increase in global temperature of the Earth below 1.5 to 2 °C (3 °F) above pre-industrial levels by limiting anthropogenic emissions. There was no such control on Venus, with its natural greenhouse emissions having increased on a vastly greater scale than we are experiencing. The planet's present state of health, a hell of extreme temperatures and dry, sterile, bare rock, is a horrific vision of what catastrophic and extreme climate change is like. A two-degree increase in global temperature for the Earth doesn't sound like much, especially when compared to the 500-degree rise suffered by Venus. But even a small step like this towards the climate of Venus could be a step too far, if not for the survival of our planet, then for the survival of our species.

CHAPTER 4

Earth: balanced equanimity

- ⊜ Scientific classification: *Terrestrial planet.*
- ⊜ Distance from the Sun: *150 million km (93 million miles).*
- ⊜ Orbital period: *1 year (365.26 days).*
- ⊜ Diameter: *12,756 km (7,926 miles).*
- ⊜ Rotation period: *1 day (24 hrs).*
- ⊜ Average surface temperature: *15 °C.*
- ⊜ Secret confession: *'I was happy with the cyanobacteria, who changed the atmosphere for the better, but those humans are too many, they are messing everything up and I am thinking about getting rid of them.'*

Earth is our home, and its character is more familiar to us than that of the other planets. Earth acts as a parent – it is sometimes referred to as Mother Earth. She provides for our needs, offering air, food, drink and shelter. As her children, we used never to question that she would continue to do so. But we grew up and realised our parent's limitations in 1968. As when we are passing out of adolescence and suddenly see our parents as people, the event, a key moment in the Space Age, was a transformation in the way that we perceive the Earth.

The turning point occurred three days into the flight of *Apollo 8*. The space capsule was in orbit around the Moon, crewed by astronauts Frank Borman (the mission commander), Jim Lovell and William 'Bill' Anders. The mission was an essential exploratory and test stage in the *Apollo* programme before the lunar landings. The astronauts orbited the Moon several times, skimming its surface, gazing down from a height of 100 kilometres (300,000 feet). They viewed the landscape in perspective, stretching out to the lunar horizon, as if they were terrestrial explorers standing on a mountain peak, looking over lands newly discovered. What they saw was not, however, an ordinary landscape. The bare grey rock was lifeless under a cloudless, black sky, and, given the Moon's lack of air, they could see clearly to the horizon without a progressive haze to provide a sense of distance.

The same view must have been available to them during previous orbits three times before. But the lunar surface below had taken all their attention because their task was to photograph potential landing sites. It was only at the beginning of the fourth orbit on Christmas Eve 1968 that they raised their heads enough to notice the Earth rise over the lunar horizon in front of them. Borman was the first to notice the earthrise: 'Wow, is that pretty!' Anders took a picture, later released by NASA under the title *Earthrise*.

Later Lovell explained how he saw the Earth: 'Up there, it's a black-and-white world. There's no color. In the whole universe, wherever we looked, the only bit of color was back on Earth ... It was the most beautiful thing there was to see in all the heavens. People down here don't realize what they have.'

Anders has also commented on the 'startlingly beautiful sight of our home planet'. As seen by the *Apollo* astronauts, the view from afar is of a blue planet, its main colour due to the harmonious interaction of blue sky and blue oceans. White polar caps show where snow lies, and ever-changing wispy white clouds show the existence of an atmosphere. Dark regions delineate the continents. The night-time hemisphere sparkles with lights on dry land where there are cities and roads connecting them, and on the sea where fishermen use lights to attract their quarry to their nets and oilmen burn off vented gases from wells. Earth as a planet shows enormous variety: it multitasks, participating in many activities.

The phrase 'blue planet', describing how it looks from space, became both the literal description of the Earth and, by extension, a metaphor for its ability to support life, including us. In a lecture called 'No Frames, No Boundaries', an *Apollo 9* astronaut Russell ('Rusty') Schweickart eloquently described an astronaut's view of the Earth as seen from the Moon:

> It is so small and so fragile and such a precious little spot in the universe that you can block it out with your thumb. And you realize that on that small spot, that little blue and white thing, is everything that means anything to you – all love, tears, joy, games, all of it on that little spot out there that you can cover with your thumb. And you realize from that perspective that you've changed, that there's something new there, that the relationship is no longer what it was.

The population of the Earth is in the *Earthrise* picture, but too small to see. We are isolated on our tiny planet in a vast Universe. As US poet Archibald MacLeish wrote, we are all on it together: 'To see the Earth as it truly is, small and blue and beautiful in that eternal silence where it floats, is to see ourselves as riders on the Earth together, brothers on that bright loveliness in the eternal cold – brothers who know now they are truly brothers.' The picture became one of the most reproduced photographs of all time.

The picture also shows that our home is limited in its capacity to support us. We are vulnerable. The photographer Galen Rowell called it 'the most influential environmental photograph ever taken'. It has been credited with helping to spark the budding environmentalist movement.

The reason we have an environment to take care of is because Earth lies in the Goldilocks Zone of the solar system. In general, the closer to its parent star (in our case the Sun) that a planet lies, the hotter it is. Close to the star, water turns to steam, evaporates and escapes into space – life dries out and dies. Far from the star, water is frozen solid into ice, and biochemical reactions are impossible – life is suspended in stasis. In an intermediate zone a planet is, like the porridge that Goldilocks found on the bears' table, not too hot, not too cold, but just right for the planet to sustain liquid water. As a consequence, on Earth there is water in abundance, in oceans that are the key to the life that has evolved there and moved onto the land, but not too far from supplies of water.

The Goldilocks Zone of a planetary system is a rather crude way to decide whether life is possible on a planet

– whether it is habitable, in other words. But there are other factors than distance of a planet from its sun. The temperature of its surface is not just a matter of how much heat the planet receives. The possession of an atmosphere, or not, is key. If the atmosphere has white clouds, they reflect heat back into space. In addition, depending on its composition, the atmosphere will, via the greenhouse effect, trap heat that does get through from the star to the planet's surface. Both are factors that play large parts in determining the temperature of Venus and the Earth.

A further key effect of the atmosphere is that it moves heat over the surface of the planet by convection of air, and winds: this tends to spread warmth evenly, ironing out differences. The Earth's personality is, as a consequence of its atmosphere, one of warm equanimity.

Even so, the temperature of the surface of a planet varies from place to place: there might be sources of heat other than the parent star – geothermal activity, for example. So even outside the Goldilocks Zone there might be localities where the temperature is between ice cold and boiling hot, and it is possible for water there to be liquid and sustain life – the moons of Jupiter (Chapter 10) and Saturn (Chapter 13) are examples.

The temperature on Earth depends on the latitude of the place in question: it is cold at the poles and warm at the equator. This is because the intensity of the Sun's heat depends on the angle of the Sun – sunlight and the Sun's heat are more intense when the surface directly faces the Sun. The details depend on the rotation of our planet and the tilt of its rotational axis.

The temperature at a given location also depends on where the planet is on its revolution around the Sun. These variations cause the progression of the seasons, an effect that is mostly a matter of how the Earth is oriented towards the Sun. If the north pole of the Earth is tilted towards the Sun, the northern hemisphere is warmer – this is summer in the northern hemisphere, winter in the southern hemisphere. As the Earth revolves around the Sun during the year, the rotational axis of the Earth stays fixed in space, so after six months it is the south pole that is tilted towards the Sun and the north pole that is pointed away – this is summer in the southern hemisphere, winter in the north. The tilt of the axis of our Earth is 23.5 degrees, a 'good' angle which helps even out the intensity of the Sun's heat over the year.

The tilt of the Earth's axis is the main cause of the annual cycle of the weather. There is an additional effect due to the eccentricity of the Earth's orbit around the Sun. The Earth is closest to the Sun in the first week of January and furthest from it in the first week of July. People who live in the northern hemisphere find it hard to believe as they shiver in the middle of winter that they are 5 million kilometres (3 million miles) nearer to the Sun than when they sweat in the middle of summer.

Over any one year, the Earth's axis points in the same direction in space and the eccentricity of the Earth's orbit remains constant. So, the effect repeats from one year to the next. However, in the longer term the orbit of the Earth changes, so the size of the effects changes. The Earth's axis wobbles with a period of 26,000 years, and the tilt does not remain the same at 23.5 degrees but nods between 21.5 and 24.5 degrees over a period of 41,000 years. The larger the tilt, the greater the variation of the seasons. This range is significant, but actually rather limited, considering how big it might be. The reason is that the gravitational pull of the Moon stabilises the wobble of the Earth and holds its rotation axis more stable than it would have been without the Moon being close by and so large – this is a property of the Moon that is good for us. The eccentricity of the Earth's orbit also changes; it varies between almost nothing and up to twice its current value over a time of 100,000 years.

In 1913, the Serbian civil engineer and geophysicist Milutin Milankovič calculated these three cyclic variations of the Earth's orbit and the way they affected the amount of sunlight falling on the Earth's surface. The variations are called Milankovitch cycles (the spelling is the conventional scientific, English phonetic representation of his name). Milankovič identified the main recent periodicity of the combination of these effects as 100,000 years.

The work of Milankovič was largely scorned or ignored for fifty years: scientists did not believe that these simple and rather small changes in the way that the solar radiation falls on the Earth's surface could affect climate. However, his work has been taken up by climatologists in the last few decades, following the discovery of some scientific evidence that the Milankovitch cycles really do affect the climate.

Temperature changes on Earth can and do affect the composition of ocean sediments and Antarctic ice cores. The sediment deposited on the seabed and the snow that falls on the Antarctic continent fall in annual layers, and the composition of each layer is a record of the temperature when it was deposited. Cores of mud and ice have been drilled out from layers of sediment and snow, several kilometres deep. Studying these cores, geophysicists can pick out the Ice Ages. For the last 3 million years, glaciers have advanced and retreated every 40,000 to 100,000 years, and the Milankovitch cycles are evident in this periodic behaviour. What Milankovič found is regarded by the National Research Council of the US National Academy of Sciences as being 'by far the clearest case of a direct effect of changing [orbital characteristics] on the lower atmosphere of Earth'.

However, climate change is complicated by changes in the behaviour of the Sun, terrestrial volcanism and continental drift, the amount of cloud cover and, in particular, changes in the composition of the Earth's atmosphere due to the emission of greenhouse gases of both anthropogenic and natural origins. Life itself manages (or disrupts) changes so that the planet stays balanced with the life that it sustains.

This principle is known as the Gaia hypothesis, formulated by the chemist James Lovelock. Gaia is the Earth goddess in Greek mythology. The hypothesis suggests that life interacts with its environment and forms a self-regulating system that maintains and develops the conditions in which life can continue. It is one of the reasons why life has evolved and lasted on Earth for such a long time in the face of major environmental changes. Just as in a family home, in which parents and children grow together and alter the function of rooms, changing the nursery into a guest bedroom, for example, the Earth and life on it have also developed over time by mutual interaction. Changes wrought by life, for example, have altered the composition of the Earth's atmosphere and its surface rocks. The changes have so far been beneficial to sustaining and developing life, although there may be a blip in progress as anthropogenic emissions work in the direction of damaging it.

The most dramatic change in the Earth caused by life was the Great Oxygenation Event of several billion years ago. It came about as follows.

The Earth was built up from rocks that gathered from the solar nebula surrounding the newly formed Sun. As these rocks settled and crushed together, they liberated the gases they carried with them. These gases formed the earliest atmosphere of the Earth, and are the same gases that remain in the giant planets like Jupiter and Saturn even now. At about 4.6 billion years ago, the principal gas was hydrogen, which joined chemically with other common elements to make water vapour (hydrogen and oxygen), methane (hydrogen and carbon) and ammonia (hydrogen and nitrogen). There were, no doubt, also the so-called noble or inert gases such as helium, neon and argon. Helium and neon are the second and fourth most abundant elements in the Universe, but they do not combine chemically with anything, so they do not get anchored to solids or liquids. They are always very light-weight gases and readily escape into space – there is now no more primordial helium left in Earth's atmosphere at all, and only slight traces of neon.

By 4 billion years ago, volcanoes and asteroids from the Late Heavy Bombardment added gaseous nitrogen and carbon dioxide into the mix. The carbon dioxide readily combined in water with the minerals of the rocks to make sedimentary carbonate minerals, and by 3.4 billion years ago the atmosphere was principally nitrogen gas. The earliest known fossils of life are stromatolites, which date from about this time, but there are chemical signatures in older rocks that suggest that life was active even earlier. Stromatolites are rocks piled up from mats of sedimentary rocks created from layers of cyanobacteria, which are single-celled microbes akin to algae: the mats are related to the floating green layers of slimy algae in the sea or on a lake known as an 'algal bloom'. Cyanobacteria photosynthesise: they take in carbon dioxide and, using sunlight, activate a chemical reaction that provides energy and body mass for the organism to live. The reaction releases free oxygen.

At first the oxygen was soaked up by the methane and ammonia in the atmosphere and other chemically active substances like iron in rocks. But, by about 2.4 billion years ago, more oxygen was being produced than was being soaked up, and the Great Oxygenation Event occurred – for the first time the atmosphere contained free oxygen. This allowed life to evolve to a second model of energy generation. Instead of taking in carbon dioxide and using photosynthesis to generate energy while releasing oxygen, animals ate and digested carbon-rich material like plants, using oxygen to generate energy, releasing carbon dioxide. Today, the Earth's atmosphere is still nitrogen-rich and retains some argon – nitrogen accounts for 78 per cent of the atmosphere and argon 0.9 per cent – but carbon dioxide has dropped to 0.04 per cent while oxygen has risen to 21 per cent.

In this way the secret life of the Earth has been inextricably interlinked with the secret life of Life. Other events that altered the course of life on this planet have not been so kind as the Great Oxygenation Event, such as the impact of large asteroids or comets. One that had a major effect on the evolution of life on our planet fell 64 million years ago into what was then a shallow sea, perhaps 100 metres (300 feet) deep, located where the Yucatán Peninsula is now, in Mexico, near the present fishing harbour of Chicxulub. It is not clear whether the impactor was an asteroid or a comet, but for brevity and clarity I am going to call it an asteroid in what follows, while keeping an open mind. The effects of the event on the Earth were at first terribly disruptive and then transformational, like the effect of a great catastrophe on a nation – like the explosion of nuclear weapons on Japan during the Second World War, perhaps, but on a larger, global scale.

The asteroid, 10 or 15 kilometres (6 to 10 miles) in diameter, took a second or so to traverse briefly through the atmosphere, far faster than the speed of sound. It thrust the air aside, leaving an empty tunnel. It compressed and heated the air at the bottom of the tunnel, and plunged into seawater, creating superheated steam in less than a second. It impacted into the seabed and shattered, taking a few seconds to pulverise and melt the rock of the seabed, excavating in minutes a crater 100 kilometres (60 miles) in diameter and 30 kilometres (100,000 feet) in depth.

The amount of material ejected from the crater was 300,000 cubic kilometres (70,000 cubic miles). It was constituted of thousands of billions of tons of rock fragments, mixed with ocean water. The fragments exploded out like shrapnel and were immediately fatal to any animals, such as dinosaurs, within range. The energy of the explosion was the same as 10 billion Hiroshima atomic bombs, equivalent to a millennium of the output of energy from Earth's entire volcanic activity. It was the kinetic energy of the motion of the asteroid that did it: the same kind of energy that crumples the bodywork of your car in a collision. Your car weighs a couple of tons, the asteroid would have weighed a million, million times more; your car might crash at 40 mph, the asteroid was moving at 40,000 mph, so altogether the impact was a million, million, million times the energy of a car crash. This is why the impact did so much more damage!

The impact created a tower of hot gases, superheated steam and glowing, molten rock, a fountain of material heated to thousands of degrees that shot up the empty tunnel in the atmosphere, quickly enveloped by a mushroom cloud of debris. Animals within sight of the hot tower were roasted.

The heat and the explosion caused the air around the site to rush outwards in a supersonic shock wave. The shock front arrived at herds of browsing dinosaurs many hundreds of kilometres away without warning. What had been silence, disturbed only by munching noises and an unusual glow of light over the horizon, turned to a cacophony of noise and tornado-like winds by which individual dinosaurs were picked up and slammed into cliff faces, and by which whole trees were uprooted and became flying clubs and lances.

Meanwhile, at the impact site, seawater had been displaced in a massive wave. The water started to refill the hole that had suddenly appeared in the sea, pouring into and propagating in and over the walls of the hole. The giant wave that resulted could have been 100 metres (300 feet) high – an enormous tsunami. Over the next hours the tsunami swept up the eastern seaboard and out into the Atlantic Ocean, a rushing flood of water, surging in and out of coastal areas. Land-dwelling creatures were drowned as the tide came in and overflowed the coastline, sea-dwelling creatures were left stranded, suffocating in the air on the exposed seabed as the tide went out. In subsequent surges, their broken, dead bodies were swept up with sand and mud and washed into layers.

In later times some of these layers became rich fossil beds. There is one exposed at Edelman Fossil Park, in a disused quarry behind a home-improvement store in Mantua, NJ. It has been excavated in a citizen science project by Ken Lacovara of Rowan University and excited students. Fossils of the broken bones and shells of prehistoric land-dwelling and marine dinosaurs, crocodilians, turtles, fish, ammonites, brachiopods, molluscs and bivalves litter the 10-centimetre-(4-inch)-thick stratum in a mass death assemblage.

If a large asteroid were to impact into the sea like this in the near future, say into the North Atlantic Ocean, the immediate damage for us would be similar. The tsunami would propagate onto the Eastern Seaboard of the USA, the coasts of northern Europe – Norway, Ireland, Britain, France and Portugal – and the more distant coasts of South America and Africa. The death toll would depend on the size of the asteroid and the location of the impact site, but could well be millions.

The tower of debris over the impact site at Chicxulub began to disperse. The debris was hot and its radiation started fires in forests worldwide. Solid pieces of rock that had been ejected into space orbited for a while and then fell back to Earth's surface in a long-lasting meteor shower.

The debris covered the world. Its material remains identifiable as a geological layer in the Earth's rocks. The layer can be distinguished from other layers and proved to be extraterrestrial, because it contains a high concentration of the element iridium. Iridium deposited by asteroids on Earth when the planet was formed has, mostly, sunk into the Earth's core, and is rare in its surface layers. Iridium-rich material must have arrived in asteroids after the formation of the Earth's core.

The iridium-rich layer from the Chicxulub asteroid constitutes the KT boundary. It divides layers of rocks from the Cretaceous geological period from layers in the Tertiary period. (KT is the accepted abbreviation in geology for Cretaceous-Tertiary, using the initial of the German word *Kreide*, 'chalk', for the rock distinctive of the Cretaceous period.)

Finely powdered debris remained suspended in the atmosphere for weeks to years, including sulphates from powdered gypsum, the mineral of the Yucatán seabed. This material blocked out the Sun, much as would happen after a wide exchange of nuclear weapons in a nuclear war, so that a 'nuclear winter' followed the firestorm. Our blue planet turned grey, with ice everywhere.

These events of 64 million years ago, perhaps together with the massive outpouring associated with the Deccan Trap volcanic formation in India, which occurred at the same time, caused a widespread extinction of many land-dwelling species, the so-called 'KT extinction event'. However, while most dinosaurs became extinct, mobile, flying, feathered dinosaurs were able to seek out the most favourable niche environments in the devastation and survived to evolve into birds. Small, seed-eating animals were able to hunker down below ground in their burrows, and survived to evolve into the mammals both large and small – rodents, bovids, primates, etc. – that dominate the land now. The Chicxulub impact was one of many random turning points, but a large one, in the life of the Earth and of the human race, an evolutionary path that eventually leads to ourselves.

The Chicxulub event was one of many impacts by asteroids and comets on the Earth. It created the second-largest meteor crater known on Earth, but there is very little trace of it readily visible in Mexico. In fact, it was discovered in 1978 by an oil prospector, Glen Penfield, in an airborne magnetic survey, which showed a curious circular arc in the seabed off the plain of agave plantations and bush near Chicxulub. There is little to be seen on the surface of the land except a shallow trough and an arc of sink holes, which mark the southern extent of the crater.

The nature of the circular features as remnants of a meteor crater was confirmed by the discovery of quartz, transformed into the minerals coesite and stishovite by the shock of the impact. These minerals are derived from silica and are dense, heavy structures akin to glass. Coesite was synthesised in 1953 by Loring Coes Jr, an industrial chemist, subjecting quartz to extremely high pressures and temperatures. It has been found in craters left by testing nuclear explosions but had never been found in any naturally occurring rock until it was discovered in 1960 in the Barringer Meteor Crater in Arizona by geologists Edward Chao and Eugene Shoemaker. Stishovite, likewise named after the person who first synthesised it, Sergei M. Stishov, a Russian physicist, is similar but formed at even higher temperatures and pressures: it is also found in the Barringer Meteor Crater.

These minerals are used as a diagnostic to identify craters of unknown origin as meteor craters. They are telltale clues, hidden in the ground, to catastrophic events in the secret life of the Earth, events lost in the past until now.

The reason why the Chicxulub crater has all but disappeared is that its walls have been eroded, its hollowed centre filled by the action of weather and the form of the land changed by the shift of the South American tectonic plate on which it is situated towards North America. The Earth is the planet on which the existence of tectonic plates is the most evident.

Tectonic plates originate in the following way. When the Earth was newly born it became molten, heated by the energy liberated as its embryonic planetesimals – asteroids – crashed down from space, and by the radioactivity of the elements in its core. Iron and similar elements then liquified and percolated down into the centre of the planet. The temperature remained high in the core, overlaid by a blanket of rock known as the mantle, which acts as a blanket, as the name suggests, but the outer layers cooled. The planet settled into its current layers, with a dense iron core and a solid rocky mantle and crust, a malleable lower mantle lying between. Further cooling caused slabs of the crust and upper mantle to become denser and sink into the lower mantle, where they floated about, a jigsaw of tectonic plates that jostled and collided. Dense slabs that collided with lighter plates rolled in jerks underneath ('subduction'), creating earthquakes, and weakened the line of collision so that molten material from below could well up in explosive volcanoes.

The collisions between tectonic plates around the Pacific Ocean are the cause of the 'Ring of Fire', the series of nearly five hundred volcanoes that stretches from New Zealand, north through the Philippines, Java and Japan, eastwards through Alaska, and south down the US Pacific coast and Mexico into Peru and Chile in South America. Colliding lighter plates buckled up like carpets sliding together on a polished floor and folded into mountain ranges like the Himalayas, the Alps, the Andes and the Rocky Mountains.

The fact that the Earth has a dense core was a secret that was uncovered in 1774 by the then Astronomer Royal Nevil Maskelyne using a Scottish mountain known as Schiehallion. The idea was to verify Isaac Newton's principle about gravity that masses attract one another. Newton himself put forward the idea of how to do this, but failed to follow it up because he thought the effect would be too small to measure. The Royal Society of London formed a 'Committee of Attraction' in order to orchestrate an attempt.

Newton imagined a pendulum, which normally hangs straight downwards in the Earth's gravitational field. However, if it were moved to be beside a mountain, the mountain would pull the pendulum off the vertical. The change of angle, as measured against the stars, could be measured, and that would give the pull of the mountain sideways to compare with the pull of the Earth downwards. Schiehallion was chosen for the experiment in 1774 because it is isolated from other mountains that could disturb the measurements, and it has steep sides so the pendulum can be close to its centre of gravity and be pulled strongly.

During his six-month expedition, Maskelyne had to fight the weather, since cloud hung about the mountain (its name refers to 'constant storms'). The cloud not only interfered with his observation of stars to establish the vertical, but also impeded the surveys he organised to determine the volume, and therefore the mass, of the mountain. These surveys were not completed until the following year. We can get a sense of the euphoria felt by the surveyors from the fact that at a drunken party organised to celebrate the end of their work they accidentally set fire to their base camp and burnt it to the ground. The measurements gave the mass of the Earth, from which its average density could be derived. Modern figures give the average density of the Earth to be 5.5 grams per cubic centimetre, compared to the average density of rock on the surface of our planet of around 3.0. There must be a much higher density core inside the Earth.

The structure of this core was uncovered by the Danish geophysicist Inge Lehmann in 1936. She studied seismic waves travelling through the Earth, which pass through the planet's central regions before being picked up by seismometers located elsewhere on the surface. Lehmann found that the core is divided into two. An inner core of iron and nickel is solid, with a diameter of 2,440 kilometres (1,520 miles), a temperature of about 6,000 °C (11,000 °F) and a density of 13 grams per cubic centimetre. It is surrounded by a shell of iron and nickel, a liquid outer core, with an outer diameter of 6,800 kilometres (4,200 miles) and a density of about 10 grams per cubic centimetre; this is a couple of thousand degrees Celsius cooler than the inner core.

Convection in the liquid outer core is driven by heat escaping from the inner core below. The circular movement of the liquid iron generates a magnetic field, much as a dynamo does; the rotation of the Earth and friction with the solid inner core play a part. The geodynamo is the origin of the Earth's magnetic field. It is more or less aligned with the spin axis of the Earth, but not exactly: at the present time the north pole of the magnetic field is in Canada. Nor is the magnetic field stable. The direction in which the magnetic field points wanders around near the Earth's poles. According to the geological record of the Earth's magnetism, which is frozen in iron-bearing rocks, the magnetic poles sometimes switch over completely.

No one knows how this all works – it is a complicated secret of Earth's life as a planet. What is clear is that the magnetic field is an essential shield for the Earth's atmosphere, protecting it against particles emitted by the Sun. The geological record of magnetism is not fine enough to be able to know how long the Earth will be without a magnetic field when the poles switch over. Years? Millennia? At that time, for a period, the Earth's atmosphere will be without a defence. We know from the fossil record, which continues across the switchovers, that this will not be catastrophic, but it may be unpleasant.

Plate tectonics will cease when the outer layers of the Earth have cooled enough to completely solidify, perhaps in a couple of billion years. This will be the end of the mountain-building era on Earth, and the high mountain ranges will gradually wear away under the processes of erosion, becoming hilly plateaux. Individual volcanoes or small volcanic ranges may happen for a while, building over weak spots, as in Hawaii and on Mars and Venus, two planets without tectonic plates. But even this activity will cease as the Earth cools further. Earth will begin to die. Eventually, even its liquid iron core will solidify and convection will seize up. Our planet's magnetic field will die away completely and permanently. Unlike the temporary loss of magnetic field during the switchovers, this permanent loss will be catastrophic. Unimpeded, the Sun's particles will scour away the atmosphere. With no air pressure to keep water molecules from escaping from seawater, the oceans will boil away, rainfall will cease, the land will dry out. Earth will have lost its equanimity and turned into Mars.

CHAPTER 5

The Moon: almost dead

- Scientific classification: *Satellite of Earth.*
- Distance from the Earth: *384,400 km (238,855 miles).*
- Orbital period: *1 month (27 days).*
- Diameter: *0.272 times Earth, 3,474 km (2,159 miles).*
- Rotation period: *Synchronous.*
- Average surface temperature: *–20 °C.*
- Secret conceit: *'On Earth they say I am dead, but I was powerful once – I used to raise mountains in minutes, not in the millions of years it took Earth to do the same.'*

By contrast with the Earth, the Moon is an airless, dusty satellite, carrying the inconsequential footprints of a dozen astronauts, their abandoned machinery, and a few dead robotic spacecraft, but very few signs of life. One of the few is the so-called Lunar Micro Ecosystem, a sealed cylinder containing seeds and insect eggs to see how plants and insects could be grown together in an artificial biosphere. If we are to colonise the Moon it will be necessary to establish some kind of garden to make food, and this was a pilot experiment towards that end. The Ecosystem was taken to the south pole of the Moon in the Chinese *Chang'e 4* lander. All started well in 2019 when the seeds were watered and sprouted, but the seedlings were killed by the low temperature of their first lunar night, when the temperature dropped to −51 °C. This does not bode well for plans to settle on the Moon.

The Moon is part of our own life. It provides light reflected from the Sun during the night-time, and in places where there is no artificial light it governs human activity, even if that activity was simply sitting in the mouth of a cave looking up at the Moon's pattern of grey markings. The Moon keeps the same face towards the Earth, so we always see the same arrangement of grey patches – they are recognisably always the same distinctive pattern, even though what the pattern is recognised as varies from culture to culture. People in Western cultures see the patches as the 'Man in the Moon', or as an Old Lady carrying firewood on her back. In Asian cultures (Chinese,

Japanese and Korean) the markings are seen as a rabbit – this is why the Chinese roving spacecraft sent to explore the Moon in 2013 in the *Chang'e* space mission was called *Yutu*, or Jade Rabbit, the pet rabbit of the Chinese Moon goddess, Chang'e. Of course, we see the pattern through the psychological phenomenon called pareidolia: we see the pattern but none really exists.

The phases of the Moon, caused by the changing relationship of the Moon, the Sun and the Earth, repeat on a regular cycle of a month, as the Moon revolves in orbit around the Earth. This gives a unit of time conveniently intermediate between the day and the year that humans have used to order their lives for thousands of years. The Ishango bone is the fibula (calf-bone) of a baboon, found by archaeologists in the Congo, used in Palaeolithic times as a knife-handle, and marked with notches as six months' record of lunar phases. Its precise date is uncertain but has been variously estimated between 6,000 and 9,000 BCE, or even earlier. The handle of the knife may have been used by a hunter to keep track of a long trip, perhaps to time his return, or to predict the movement of game during moonlight. Alternatively, knives are common in household activities, which suggests it may have belonged to a woman, and the knife-handle may have been used by her to keep track of her fertility.

The Moon is also responsible for the tides, pulling the sea onto and away from the shore, through which it influences the life of sea creatures, dictating their time to eat, to grow and to reproduce, as well as affecting our ability to travel on the sea's surface.

Like the planet Mercury, the Moon is pitted with impact craters, most of which were made in the two bombardments that happened as the solar system began. Asteroids that have arrived most recently tend to be smaller, so they make smaller craters, which pepper the older plains.

Between 1969 and 1972, the Apollo astronauts left a total of six seismometers on the Moon, to look for 'moonquakes'. They are the equivalent to medical monitoring equipment in the high dependency unit of a hospital to check the vital signs of a patient, revealing what is going on unseen inside the body. The seismometers operated until 1977 and recorded hundreds of small events.

Some events came from the deep interior of the Moon, and analysis showed how the Moon's core is layered, with an inner solid core surrounded by a mantle. Quakes are excited in the core by tidal forces due to the Earth, not by plate tectonics.

Some moonquakes come from the surface layers, shocks generated when rocks emerge from the cold lunar night and are suddenly exposed to the Sun's heat, so they expand, and there is a shock when the resultant strains are suddenly released. Still others come from occasional meteor impacts.

Both NASA and ESA monitor the night-time areas of the Moon to watch for impacts. Every few hours they see a brief flash of light that is the signal that an impact has happened – extrapolating to account for the unmonitored areas and the flashes that are missed, the rate scales up to about eight impacts per hour on the whole Moon. These flashes come from the impact of meteors that are a few kilograms in mass, and each probably leaves a crater a few metres in diameter, which is below the size that is readily visible. Larger craters are still occasionally being made. The *Lunar Reconnaissance Orbiter* has accumulated multiple and repeated pictures of the lunar surface since its launch in 2009, enabling changes to be picked out. Hundreds of new craters with diameters over 10 metres have appeared, at the rate of 180 per year.

These impacts help to churn over the top soil of the Moon, a process referred to as 'gardening'. The top two centimetres of the lunar soil is turned over every 100,000 years. The impacts will be something that will have to be taken into account if a permanent colony is established on the Moon, which is why NASA and ESA are particularly interested in the issue. The Moon is not as vibrant and active a world as the Earth, but it is not the eternal, unchanging, dead place it is sometimes said to be.

Meteor impacts have covered the Moon with dust. There is an unforgettable, much-reproduced photograph made in 1969 by astronaut Buzz Aldrin, the second man on the Moon, of his boot print in the dust. It showed NASA engineers back in Houston the depth and consistency of the lunar soil, to help them design the wheels of lunar buggies for good traction. The value of the photograph extends beyond the technical. It marked in a poetic way the moment and the place where humans first landed on another world.

The landing site for *Apollo 11* was in Mare Tranquillitatis (Sea of Tranquillity), a dusty plain of basaltic lava. It is smooth, with few craters and boulders, a safe landing area, and no large hills, high cliffs or deep craters. The view of the lunar landscape from Tranquillity Base (as the landing site was called) was strongly illuminated by the remorseless light of the Sun shining out from a black sky from beside the Full Earth – because there is no atmosphere, there is no air and no blue sky. The shadow of the lunar module *Eagle* was sharp and intense. The terrain was very flat. Buzz Aldrin described the view from the windows of *Eagle*:

We'll get to the details of what's around here, but it looks like a collection of just about every variety of shape, angularity, granularity, about every variety of rock you could find. The colors – well, it varies pretty much depending on how you're looking relative to [the direction opposite the Sun]. There doesn't appear to be too much of a general color at all. However, it looks as though some of the rocks and boulders, of which there are quite a few in the near area, it looks as though they're going to have some interesting colors to them . . .

Armstrong told Mission Control that

The area out the left hand window is a relatively level plain cratered with a fairly large number of craters of the 5 to 50 foot variety, and some ridges – small, 20, 30 feet high, I would guess, and literally thousands of little 1 and 2 foot craters around the area. We see some angular blocks out several hundred feet in front of us that are probably 2 feet in size and have angular edges. There is a hill in view, just about on the ground track ahead of us.

Determined to provide fitting first words from the Moon, Neil Armstrong concentrated on the historic moment: 'That's one small step for a man, one giant step for mankind.' Looking around, he described the landscape: 'It has a stark beauty all its own. It's like much of the high desert of the United States. It's different but it's very pretty out here.'

Buzz Aldrin agreed:

In every direction I could see detailed characteristics of the gray ash-colored lunar scenery, pocked with thousands of little craters and with every variety and shape of rock . . . With no atmosphere there was no haze on the moon. It was crystal clear. 'Beautiful view,' I said . . . I slowly allowed my eyes to drink in the unusual majesty of the moon. In its starkness and monochromatic hues, it was indeed beautiful. But it was a different sort of beauty than I had ever seen before . . . 'Magnificent desolation'.

Apollo 12 landed in an area of the Moon that was much the same as Tranquillity Base. The *Apollo 13* mission had to be aborted. *Apollo 14* landed near Cone Crater, and the astronauts wanted to look down into it, but got lost and had to turn back to base as their supplies were running low. The *Apollo 15* landing site was the most adventurous. The lunar module *Falcon* landed in 1971 on a dark plain near the Montes Apenninus, with Mt Hadley the nearest peak. During the three days that they spent on the lunar surface, they drove the Moon Buggy twice to the Hadley Rille, a valley along which molten lava has flowed, deepening the channel, lining it with solidified lava and making it steep sided. They discussed whether to go into it and descend to the valley floor, and sensibly decided not to risk doing so:

> I could see the bottom itself – very smooth, about 200 metres wide, and with two very large boulders right on the surface of the bottom. 'It looks like we could drive down to the bottom here on this side, doesn't it?' Dave asked hopefully. And he actually wriggled over and found a smooth place that sloped from St George's Crater, into a gully that dropped to the bottom of the rille. 'Let's drive down there and sample some rocks.' 'Dave, you are free to go ahead. I'll wait right here for you,' I told him. I reasoned that we might have made it down and back, but if we had driven to the bottom, and something had happened to the machine, we'd never have been able to get out.

The Moon's craters are nearly circular, like all meteor craters. It does not matter much that the impactors arrived at various angles to the lunar surface. The craters are not made as the asteroid pushes aside the lunar material, which would make an elliptical hole. They are made by the vaporisation below the surface of the impactor and the rocks that it strikes. The resultant gas expands and breaks out of the surface to make a symmetrical explosion, which pushes the surrounding surface rocks out upwards and sideways.

The surrounding rocks are pulverised into chips and dust. Unimpeded by any air, the debris arcs up and over. The debris can leave trails that radiate from the crater. Some big impacts, like the one that made the crater Tycho, have made streaks that extend to the far side of the Moon and even circumnavigate it, coming back to this side of the Moon. Tycho is visible with the naked eye as the brightest spot on the Moon.

The smaller lunar craters are simple, like a bowl for breakfast cereal. Lunar craters over 15 kilometres (10 miles) in diameter are more complex, with a central peak, or even an internal mountainous ring, where the surface rock bounced up and down a few times. Some large craters have terraces on the inside of the main crater walls where the slopes of the walls have slumped back from where they were piled up.

If an asteroid lands by chance on the wall of an older crater, it makes a new, overlapping, fresher-looking crater. Apart from cases where the walls are broken by a fresh impact, the walls of the older craters have eroded somewhat, not through weather but by the repeated heating and cooling of the lunar rocks over the month as the Moon rotates. The temperature on the Moon in the lunar daytime is about 100 °C (210 °F), at night as cold as −170 °C (−270 °F). This wide variation of temperature causes the rocks to expand and contract a lot, which causes small moonquakes, and breaks off flakes and bits of dust. This process adds to the dust that is generated when craters are made by impacts and the impacting asteroid and lunar rock are pulverised.

Some of the lunar craters are huge. The largest is the South Pole-Aitken Basin, which is 2,500 kilometres (1,550 miles) in diameter, say one-quarter of the area of the contiguous United States. It spans the south pole of the Moon. Most of it is out of sight on the far side of the Moon. The crater floor lies about 13 kilometres (43,000 feet) below the crests of the mountains at the crater rim. It is the biggest hole in the solar system.

The largest impact crater on the near side of the Moon is the Mare Imbrium, the so-called Sea of Showers, which is 1,145 kilometres (710 miles) in diameter. It is visible to the naked eye as the second largest of the dark grey markings called *maria*. Both the Aitken Basin and the Mare Imbrium basin were formed a long time ago. The age of the Aitken Basin has been estimated by looking at the number of smaller craters that pit its interior, so it is not very precise. But the Mare Imbrium has been dated precisely. *Apollo 15* astronauts brought back rocks from there, which have been dated as 3.9 billion years old.

The asteroid that impacted the Moon to make Mare Imbrium was about 250 kilometres (150 miles) in diameter. Mare Imbrium consists of a flat interior area bounded by three rings of mountains, broken into a number of ranges, which have been given names that hark back to terrestrial mountain ranges. The outer ring, which we can call the crater wall, is made up of Montes Caucasus, Montes Apenninus and Montes Carpatus. The middle ring of mountains is Montes Alpes. There is an innermost ring, with a diameter of 600 kilometres (370 miles), but it has been mostly buried by the lava flow that flowed into the crater, either as an immediate consequence of the impact or at a later date, and shows only as some low hills and ridges. The various concentric features of the crater were formed as the lunar surface oscillated after the impact. Seismic waves from the impact created a chaotic terrain on the far side of the Moon (like the weird terrain on Mercury, made by the Caloris Planitia impact), and split the lunar surface with fault lines. Had anyone been anywhere on the Moon at that time, they would have felt the impact as the Moon resonated.

The floor of the Mare Imbrium crater lies 12 kilometres (39,000 feet) below the peaks of the crater wall. Beyond the crater walls is a strewn field of debris flung out of the crater, and a number of radial grooves in the surrounding terrain that have been scoured by the out-flung material, like cannon balls charging through the masts and deck structures of a wooden battleship.

On Earth, mountain ranges like the Apennines and the Alps grow millimetre by millimetre from the slow collision of tectonic plates, reaching their height of several kilometres (tens of thousands of feet) after millions of years. On the Moon, the Montes Apenninus and the Montes Alpes shot up to similar heights in a frenzy of a few minutes.

There is evidence of volcanic activity on the Moon, but it is minor compared to the scars left by the bombardment of asteroids. Besides the lava plains that fill some of the craters, the visible traces include the so-called 'sinuous rilles'. These are like rills on Earth, which are channels made by rivers. Rilles on the Moon used to be thought to be the same (astronomers continue to use the archaic spelling left over from the time when they were first seen). But they are channels which, it seems, were made when lava oozed up from below the surface of the Moon and flowed out in streams, responding to changes in the height of the surface, by winding in the way that rivers of water do on Earth. The lava drained back out of the rilles to create the steep-sided channels that are visible today. One secret that astronauts expect or at least hope to find when they go back to the Moon in the future and are able to rove around more freely is that some of the rilles may continue beyond their apparent termination, plunging below the surface as a 'lava tube'. Lava tubes are tunnels in which lava once flowed, the surface of the lava having cooled to make a solid roof before the lava subsided. These lava tubes may in the future serve as tunnels suitable for astronauts to live in, if and when they establish a lunar colony.

Almost a twin planet, it seems that the Earth–Moon system originated from the collision of the proto-Earth, Gaia, with another proto-planet, Theia, soon after the formation of the solar system – perhaps 100 million years afterwards. Gaia was 90 per cent of the size of the Earth and Theia the size of Mars.

The collision was a glancing blow, and left the Earth rotating once every five hours. The Moon ended up orbiting the Earth much closer than it does today. The tidal forces between the two bodies locked the Moon so that the same hemisphere faces the Earth; the dissipation of energy by the tidal forces acting over billions of years took energy from the orbit of the Moon and the rotation of the Earth. The Moon retreated to its present distance: it is still retreating at 4 centimetres (2 inches) per year. The Earth slowed down, the day lengthening from five hours to its present-day value of twenty-four hours: it is still slowing down.

Both proto-bodies had a core–mantle structure. The two iron cores merged into one, like raindrops that run together as they drain down a window pane. The Earth ended up with the extra-large core, and the Moon with almost none. The size of the Earth's core meant that it has remained liquid for a long time, sustaining our magnetic field and protecting our atmosphere, enabling it to continue to contribute its life-sustaining properties to our evolution. The mantle materials jumbled up and were distributed between the two bodies. As a result, the composition of the lunar rocks is essentially the same as the composition of the Earth's mantle.

If life had started on Earth before this impact, the impact would have reset the clock, because the impact would have heated the Earth and Moon to perhaps 1,000 °C (2,000 °F). It would have vaporised any liquid water present on the Earth: the water that is here now must have been released by the volcanic activity that followed the impact or have been brought here by subsequent smaller impacts from asteroids and comets. But, having potentially had a negative effect on the start of life here, the collision then had a positive effect on its development. Earth was provided with seasons by the Moon, which created the tilt of the Earth's axis at 23.5 degrees, which in turn created the stimulating variety of climates over the Earth's surface. At the same time, by making the

Moon so large, the Theia event stopped the nodding motion of the axis of rotation of the Earth from going too far, and making seasonal changes too extreme. The cycle of the seasons here on Earth – the colourful poetry of spring, the icy grip of winter, the torrential rain of the monsoon, the sweltering heat of the sirocco wind – can be traced back to this unique event.

CHAPTER 6

Mars: the warlike planet

- ◉ Scientific classification: *Terrestrial planet.*
- ◉ Distance from the Sun: *1.52 times the Earth–Sun distance, 227.9 million km (141.6 million miles).*
- ◉ Orbital period: *687 days.*
- ◉ Diameter: *0.532 times Earth, 6,792 km (4,221 miles).*
- ◉ Rotation period: *24.6 hrs.*
- ◉ Average surface temperature: *–65 °C.*
- ◉ Secret apprehension: *'I quite like being tickled when the space probes parachute down, but I am not looking forward to it when my moon, Phobos, lands.'*

Mars is the Red Planet – as red, it is said, as blood, a colour in keeping with the God of War after whom the planet is named. It is the planet whose name has been given to the most famous movement of Gustav Holst's orchestral suite *The Planets*, which describes in musical terms their characters, based on their personalities as described by astrology. Holst had been interested in astrology for a long time, and the interest was enhanced when, in 1913, he went on a holiday in Spain with Clifford Bax, the brother of the composer Arnold Bax. Clifford Bax was an astrologer, and taught its technicalities to Holst. It became his 'pet vice' and, thereafter, Holst would delight to cast horoscopes for his friends.

The opening movement of *The Planets* suite is called 'Mars, the Bringer of War', and its insistent staccato chords evoke a frightening image of mechanised warfare, tanks having been used in battle for the first time in the First World War, which was taking place as Holst sat down to write the music. The planet is easily recognisable: there is nothing else in the sky that shines with that colour and that brightness, save the star Antares, whose very name means 'rival of Mars'. The soil on the surface of Mars is what gives the planet its colour – it is made primarily of a mineral akin to the red rust that forms on wet iron and steel.

Mars is the most Earth-like planet of the solar system. It is considerably smaller, half the diameter of the Earth. It is much colder and drier, but it started its life in the same way as the Earth did, 4.6 billion years ago, warm and wet, with a thick atmosphere and abundant water. If there had been people on Earth at that time, they would have looked up into the sky and seen it, probably, as a blue planet, not a red one. But everything changed about 4 billion years ago, when things took a turn for the worse. Like Venus, Mars suffered a global climatic catastrophe, the reason for which lies, until recently a secret, within its core. But it still has an atmosphere and its landscape has a terrestrial appearance.

The true nature of the surface of Mars emerged in the Space Age. The scientific advances were hard won. More than half the space missions launched to Mars in the half-century that started in 1960 failed to complete their missions. Mars gained a reputation among space scientists for fighting back, for keeping its secrets. Good launch opportunities recur every two years, which is an irritatingly long time to wait for a second shot (and pay the mission team while they wait). A high proportion of the cost of a mission is in the development of the spacecraft, and you have to make a second one anyway, as a spare, in case there is an unfortunate accident with the first, so space missions were often launched in pairs.

Missions intended to land on a planet with an atmosphere can use parachutes to float down, but the characteristics of the Martian atmosphere are very variable over time and over the planet: it is difficult to predict several years in advance, during the design phase of the mission, what the vehicle will find when it gets there. If the atmosphere is thinner than usual, for example the air is warmer than expected, the parachutes might not brake the rate of descent enough, and the vehicle might impact the surface with too much of a bang. The mission trajectory has to be just right: the angle of entry into the atmosphere is critical. Too steep and the vehicle plunges too fast and crashlands; too oblique and the vehicle bounces off the atmosphere back into space. It is tricky.

Of course, although critical, the descent and landing are just the end of the journey: the vehicle has to survive the launch and the journey itself, with mechanisms and electronic equipment intact. The launch might go wrong and the rocket explode, the vibrations of the launch might be too violent and shake the equipment so that it fractures, or the spacecraft might be injected into the wrong trajectory. On the journey, the equipment might be affected by the space environment, whether vacuum, radiation or meteor impacts. In a vacuum the fabric of a parachute might lose its suppleness and rip, a hub and axle in a mechanism might be welded together by cosmic radiation and become immobilised, or a piece of meteor travelling at a relative speed of tens of thousands of miles per hour might hit and wreak colossal damage.

If a mission does fail, it is sad to see the downcast faces of space scientists in the control room as they realise that their work is lost – not only the work they have already done to build the spacecraft, but also the work they would have done during their scientific investigations. It could be career-changing, especially for a post-graduate student: a PhD thesis saying 'I built some equipment, and here are ten chapters about it, but I never got a chance to use it' does not provide as much weight as a thesis in which one chapter is devoted to the equipment and nine are packed full of results about the surface of another planet.

I watched the team that made the British *Beagle 2* lander as it descended to Mars on Christmas Day 2003 and failed to make contact from the surface. They clung obstinately with grim faces to a forlorn hope that they could identify the problem, find a way round it and activate the spacecraft. It seems that it landed OK, but its solar panels failed to open fully and it could not deploy its radio antenna. The atmosphere of Mars was hotter than anticipated and less dense, so maybe it hit the ground at too high a speed, and was damaged.

What the *Beagle* space scientists were hoping to do was to join the investigations as to whether there is life on Mars. This is something that is hard to do by viewing the planet through a telescope from Earth, although telescopes that

117

amateur astronomers have nowadays readily show the features on the disc of Mars. The Italian physicist and astronomer Galileo was the first to view the planet through a small telescope in 1610, but his telescope was too small and its optics too blurry for him to be able to distinguish any surface markings. The first surface feature to be described was discovered in 1659 by the Dutch astronomer Christiaan Huygens. He saw and sketched a dark triangular marking, which he described as being like a large bog. Similar grey patches on our Moon were at the time thought to be oceans, or *maria*, and he was following a similar line of thought. In fact, both kinds of features are rocks and minerals coloured differently from the rest of the surface.

By following repeated appearances of this triangular feature, Huygens was able to determine the rotation period of Mars, its 'day'. A Martian day is called a 'sol'. It is just thirty-seven minutes longer than the Earth's twenty-four hours.

Mars is twice as far from the Sun as the Earth, and its 'year' is considerably longer – 687 earth-days, compared to Earth's 365 earth-days. Its tilt in its orbit is similar to Earth's, 25.2 degrees, compared to Earth's 23.5 degrees. As a result of all this, Mars has day–night cycles and seasons like the Earth's, but the winters are much colder, and they last longer. The orbit of Mars changes with time, and the seasons and climatic cycles change by large amounts accordingly. Its surface temperature currently ranges from 20 °C (70 °F) during the day to −140 °C (−220 °F) at night. The coldest place on Earth is Dome Fuji, an ice dome on the high plateau in Antarctica, where the temperature drops down to −80 or −90 °C (−120 °F).

In 1666, the Italian-French astronomer Giovanni Cassini discovered that Mars has polar caps; they show in a telescope as white patches at the poles of Mars, one growing in the winter as the one in the opposite hemisphere withers in the summer. The polar caps are deposits of ice and dry ice, 2 to 3 kilometres (6,000 to 10,000 feet) thick, with precipitous ice cliffs at their edges. They are surrounded by plains with drifts of ice that melts in the spring. More precisely, it sublimates. 'Sublimation' is the process by which a solid turns into a gas without passing through the liquid state. 'Dry ice' is solid carbon dioxide ice that sublimates directly into gas; this is what produces a smoky or foggy effect on stage during theatrical productions. Water ice normally melts into liquid water before turning into gas ('steam') but it sublimates directly into gas if the atmospheric pressure is low, as it is on Mars. As the ice loses its grip on the land in the spring, it lets go of the soil that lies on the slopes, and the soil slides downhill in a landslip. As the landslip cascades down onto the plain below, red dust billows up into the air. The landslips can be seen from space, although of course their thunderous roar cannot be heard, since sound does not travel in a vacuum.

The atmosphere of Mars is much thinner than Earth's and made of carbon dioxide, nitrogen and argon. This is much like the composition of the Earth's atmosphere, except that there is no oxygen, which on Earth originates from vegetation, of which there is none on Mars. The Martian air is thin but it supports clouds. Some of them are large enough to be seen from the Earth. Sometimes, they can be seen trailing back down-wind from the top of a mountain. You can occasionally see a similar phenomenon at the tops of skyscrapers, or at the tips of the wings of a jet aircraft.

In 1840, the German banker and amateur astronomer Wilhelm Beer and his colleague Johann von Mädler made the first maps of Mars, showing dark areas that stayed fixed in position. Initially these were described as wet areas. They seemed variable in colour and intensity, and the French astronomer Emmanuel Liais suggested in 1860 that they were vegetation. He thought that the changes could be seasonal variations. Eventually they proved to be changes of visibility due to dust storms. The Italian astronomer Giovanni Schiaparelli mapped Mars in 1877, and labelled the dark shapes as 'continents', 'islands' and 'bays'. He thought he saw numerous long, straight *canali* (channels) connecting some of the geographical features.

The US businessman Percival Lowell used his riches to set up an observatory in Flagstaff, Arizona, to search for Pluto and to study Mars. He saw the changes in the colour of the dark areas, and correlated them with changes in the polar caps. As a polar cap reduced in size, a dark belt of blue-green fringes its outer edge, retreating as the polar cap shrinks further. A wave of darkness rolled down the *canali* towards the equator. It looked as if a supply of water was reinvigorating vegetation, much like the seasonal flooding of the Nile valley. For Lowell, a combination of a lack of familiarity with the subtleties of the Italian language and wishful thinking about life on Mars caused him to interpret the *canali* as 'canals' and the patches where they crossed as oases. He envisaged the straight lines as vegetation growing alongside artificial links connecting bodies of water, much like the green and verdant zone alongside the River Nile, with oases of vegetation. The canals, he thought, had been set up by Martians as an irrigation system to mitigate against perpetual drought, carrying water from the polar caps. The belief took hold that Mars is an old world, drying out, its inhabitants looking to colonise Earth because their own planet is dying.

This idea was given legendary form by H.G. Wells in his novel of 1898, *War of the Worlds*. Its opening lines are spine-chilling, as spoken by Richard Burton in Jeff Wayne's musical version of 1978:

> No one would have believed in the last years of the nineteenth century that this world was being watched keenly and closely by intelligences greater than man's and yet as mortal as his own; that as men busied themselves about their various concerns they were scrutinised and studied, perhaps almost as narrowly as a man with a microscope might scrutinise the transient creatures that swarm and multiply in a drop of water.

Wells's picture of Mars is riveting, but fictional. Turkish-born French astronomer Eugenios Antoniadi concluded from viewing Mars through a large telescope in Meudon, outside Paris, in 1909, that the 'canals' were psychological interpretations of faint, blotchy structures on Mars seen through the wobbly terrestrial atmosphere and not real. Alas, Mars probes in the Space Age have confirmed that there are no canals!

What was needed to establish the true nature of Mars was a close-up inspection. The first successful mission to visit the planet was a fly-by. This was *Mariner 4*, which flew past Mars on 14–15 July 1965, recording data on a tape recorder that relayed it to Earth at leisure when the fly-by was complete. Pictures of the land below the flight-path showed a surface that was heavily cratered. What *Mariner 4* saw, however, was not typical: it was, by chance, flying over one of the oldest terrains on the planet. Mars has no tectonic plates and its atmosphere is thin, so that craters formed by impacts on the oldest surfaces, up to 3.8 billion years old, have never been erased by erosion.

The first spacecraft to enter into orbit around Mars – and in fact the first to do so around a planet other than the Earth – was *Mariner 9*, arriving on 14 November 1971. It arrived during a planet-wide dust storm, the surface totally obscured. All that the space scientists could see was a featureless cloud. There were long faces in the control room at the bad luck.

Dust storms on Mars are common. They are often small, spiralling tornados of dust, which, when they occur on Earth, are called dust-devils, or willy-willies. Some native peoples on Earth regard dust-devils as spirits. They skip erratically from the top of one mound of earth to the next, like a fleeing antelope or some other agile creature. One passed me within yards when I was walking near the telescopes of the South African Astronomical Observatory in the Karoo desert in South Africa. What it brought to mind was a picture in a book of stories from *The Arabian Nights* that I had as a child, depicting the story of Aladdin. The dust-devil towered over me, just as the Genie of the Lamp dwarfed Aladdin in the picture. It was broad at the top, like the genie's head and shoulders, and it spiralled and narrowed to a wisp at its base, like the genie as it came out of the lamp. I heard it hiss as it passed. I can readily understand why dust-devils might be thought to be living and rather menacing.

Martian dust-devils roam over the Martian desert, travelling at speed and changing direction in a way that looks at once purposeful and demented. They disturb the surface layers. This exposes dark-coloured material under the red surface dust. The darker material looks from space like a scribble in the desert, marking the dust-devil's track.

Martian dust storms can be much larger than individual tornados. They can be so large they can sometimes be seen from Earth, obscuring the surface features, and they can persist for months. They kick up light, sandy dust, and, seen from the surface of Mars, they completely obscure the Sun. The dust might drift over the solar panels of a rover travelling the surface, reducing the power available to drive the vehicle onwards, bringing it to a halt. If the wind blows the dust off again promptly, the rover may revive as the Sun tops up the batteries. But if the dust storm lasts a long time, and the batteries discharge, the rover may suffer from the cold. This happened to the *Opportunity* rover, which fell silent after planet-wide dust storms in 2018. It seems unlikely that it will wake up, and the NASA scientists have had to let it go and work on the next thing.

The winds blow dust everywhere, covering the entire surface of the planet, even the ice caps, which take on a layered structure, like a chocolate-and-cream layer-cake – a layer of dust overlying the ice, the ice reforming and getting covered again. Strong dust storms happen in summer, when convection currents cause high winds that pick up the dust. The strongest dust storms happen in summer in the planet's southern hemisphere. The reason for this is that, as it happens, Mars is considerably closer to the Sun during the summer of Mars's southern hemisphere than the summer of Mars's northern hemisphere. Southern summer is therefore warmer than northern summer and convection currents are strongest at that time, so dust storms are more potent. Once started, the bigger storms can last weeks to months.

Space scientists do not give up easily. Having got to Mars at a time when they could see nothing, *Mariner 9*'s controllers postponed their exploration of the surface of Mars, literally waiting for the dust to settle, which it did two months later in mid-January 1972.

Their patience was rewarded in abundance, as *Mariner 9* discovered variety on the surface of Mars. It saw recently formed volcanoes. The largest is Olympus Mons, 24,000

metres high. Think of Mount Everest, the highest mountain on Earth at 24,000 feet high: metres are just over
three times as long as feet, so Olympus Mons is more than
three times higher. The biggest volcano on Earth is Mauna
Loa, which comprises just about the whole of the Big
Island of Hawaii, rooted deep in the Pacific Ocean below:
Olympus Mons is 100 times its volume. Its size is the
reason why this volcano on Mars was given such a grand
name: on Earth, Mount Olympus was the abode of the
gods. Olympus Mons is associated with a number of other
volcanoes concentrated in a volcanic region of Mars called
Tharsis Province. There are what look like recent lava
flows there, perhaps only a few million years old, but no
active eruptions or lava flows have ever been seen, and it
seems there is no violent seismic activity – no 'marsquakes'.

In 1972, *Mariner 9* also discovered a huge canyon
system, a rift valley, the Valles Marineris, named after the
satellite. The canyon extends 4,000 kilometres (2,500
miles) east–west along the equator. It is 600 kilometres
(370 miles) wide and 7 kilometres (23,000 feet) deep.
Think of the Grand Canyon in Arizona: Valles Marineris
is five to ten times bigger in every dimension. But, *Mariner 9*
showed that vast, dry, red or yellow plains are the most
common terrain on Mars.

The breakthrough missions that uncovered much of
the secret life of Mars were the *Viking* missions of

1975–6. There were two missions. Each comprised two parts: a lander, and a mother ship that stayed in orbit around Mars after dropping it. The two landers were the first to touch down on the surface of Mars. *Viking Lander 1* remained operational for six years, *Viking Lander 2* for three. They looked for but failed to discover biochemistry – the surface was sterile. There seems not to be life on the surface of Mars, at least not life that had the properties tested for.

The big surprise was the evidence provided by these two missions that vast amounts of water once covered large areas of Mars. The evidence was in the form of geological structures normally made by water. There were very flat areas lying within closed basins – evidently consolidated silt that accumulated under lakes. The lakes formed layers of clay. Clay is a word that is rather loosely used in common language but in geology it has a technical meaning that goes beyond its appearance. The flat layers were investigated later by downward-looking satellites to determine their chemical composition. Their chemical structure was proved definitely to be minerals deposited by water, confirming that parts of Mars were once under standing water.

The *Viking* programme saw valley systems on Mars, with small, meandering valleys joining larger valleys at a lower level – evidently dried stream- and river-beds.

Some valley systems had characteristics that suggested that the streams and rivers had run on the land surface under ice sheets or glaciers. Yet further evidence was in the form of tear-drop-shaped 'islands' standing above the plains, downstream of obstructions like the walls of craters. The shores of the now marooned islands were in the form of cliffs hundreds of metres high, swept and eroded by a surging flood. Some plains were littered with rounded boulders, once tumbled in fast-running water. The evidence grew that, in the past, Mars had had abundant water, in a time that has come to be known as the Noachian era, the adjective referring to Noah and the biblical Flood.

Water is a prerequisite for life. Life on Earth originated in the oceans, perhaps deep down near sea-floor volcanic activity. Oceanic creatures may have crept onto the sea-shore and now dwell on land, but they need to drink to replenish the water lost through their biological processes. Water is necessary as a solvent for the biochemical activity that makes life work. Finding evidence of massive amounts of past water on Mars created a suggestion that there might be small quantities of water there still, and was an impetus to look more widely and more carefully for evidence of life.

The pace of Mars exploration picked up from the year 1980 – fewer launches but longer-lasting missions, with much more being discovered. Since 2010 there have always been four to eight spacecraft active on and around Mars. The most spectacular vehicles are those that have landed on Mars and travelled the surface. These rovers have ranged in size from a small coffee table to a golf buggy, and were able to travel a total distance of 100 metres/yards to tens of kilometres/miles. The most recent rovers are able to select an interesting place to visit, through communication with Earth, and to make their way there, in part autonomously, working their independent way around obstructions. The time for radio signals to travel to or from Mars is up to twenty-five minutes, so it could easily be an hour for the rover to see a problem obstacle and receive a command from Earth on how to go around it: far quicker to make the decision locally! The studies being carried out include surface mapping, analysis of the composition of rocks, and studies of the Martian atmosphere and magnetic field.

The red colour of Mars is due to the layer of dust that covers the planet. It pervades the atmosphere and colours the sky orange. The dust is made up of various forms of haematite – a red or orange ferric oxide mineral akin to rust. This mineral typically forms in water, and was seen in abundance in the Terra Meridiani area of Mars. The *Opportunity Rover* was landed there in 2004 to investigate, with the thought that as there had been lots of water there, some evidence for Martian life might have been left behind. *Opportunity* found areas covered with small spheres made of a type of water-formed haematite that is less red than usual. When pictures taken by *Opportunity* were processed to emphasise these spherules, 'less red' became exaggerated to 'blue'; the spherules were called 'blueberries' by the *Opportunity* mission team. The claim that there are blueberries on Mars is not a statement about the discovery of life there!

The magnetic field of Mars is weak, typically less than 1 per cent of the strength of the Earth's. It is very weak over the new crust of the northern lowlands, and also over large, deep craters and active volcanic areas. It is higher in those parts of the southern highlands that are old areas, undisturbed by giant impacts or volcanism.

The magnetic field of the Earth is not something that impinges strongly on everyday life here, and one might be forgiven for dismissing the magnetic field of a planet as unimportant. But the weakness of the magnetic field of Mars is the reason why the planet changed from wet and warm to dry and cold, and may explain why life never got started on Mars or, if it did, why it never exploded to become a dominant feature of the planet, as it did here on Earth.

Our magnetic field arises mainly from an internal dynamo, caused by circulatory motions in the molten iron core of the Earth. Surrounding the Earth and its atmosphere, it extends out into space even beyond the orbit of the Moon at 400,000 kilometres (250,000 miles). The volume containing this magnetic field is called the magnetosphere. Mars once had such a magnetosphere, generated by a dynamo, which magnetised its older rocks (residual traces of magnetism in its old rocks are how we know this). But the dynamo shut down.

The reason the dynamo shut down was that the circulatory motions in the liquid iron core of Mars ceased. Why that occurred is a secret that Mars retains still. Maybe, because it is small, the core cooled quickly and became gooey and then solid. Maybe the internal structure of Mars is different from Earth's, and whatever mechanism creates the circulatory motions inside Earth does not exist in Mars. One good thing: Earth's iron core is larger and has more radioactive material in it than that of Mars, and the surface of the core, through which the core cools, is proportionately smaller. Although the core is cooling, it will take many billions of years for it to solidify, and is not something for our lifetime – one thing fewer for us to worry about!

The magnetic field around the older rocks is the weak, residual, magnetic field of Mars. The rocks in which the magnetic field is anchored formed on Mars before the dynamo shut down. Similar rocks were originally magnetised but lost their magnetism if they were heated by volcanism or meteor impacts (to temperatures above a few hundred degrees) and melted, then refroze. This is why there is no magnetic field trapped in the rocks on the floor of the Hellas basin, a large meteor crater in the southern hemisphere.

Rocks recently formed, after the dynamo shut down, have never had a magnetic field. Those in the northern hemisphere of Mars are generally younger – it is flatter and lower than the hilly southern hemisphere, apparently having been formed from recent volcanic flows and sediments. Whatever the reason, the rocks of the northern hemisphere are recent, so the magnetic field is particularly weak over the northern half of the planet.

The effect of the weakening of the magnetic field on Mars when the core froze was life-changing for the planet. The Earth's magnetosphere extends out beyond the Moon and shields our atmosphere from the solar wind, a gusty stream of electrically charged particles that emanate from the Sun. By contrast, Mars lacks this protection. Its magnetic field is weak and extends over the southern hemisphere only up to a height of, say, 1,500 kilometres (1,000 miles). At best, its magnetospheric shield becomes strong enough to form some sort of defence only close to the planet's surface. As a result, electrically charged particles in the solar wind interact with the atmosphere, and heat it. Atmospheric molecules are driven away from Mars and blown out into space at speeds in excess of 400 kilometres per second (300 miles per second).

With its atmosphere stripped away, the surface of Mars is exposed to ultraviolet light and solar particle radiation, which would be deadly to surface-dwelling life. The weak atmosphere also means that the pressure is 1 per cent of the Earth's atmospheric pressure, too low for liquid water to exist in the open, and the weak greenhouse effect means that there are severe frosts at night in the polar regions where the temperature falls to −140 °C (−220 °F).

The biography of the planet that emerges from these discoveries is that Mars was once wet and warm, with lakes and flooded craters. It had a dense atmosphere and plenty of liquid water and ice. Water pooled on the surface and formed flat lake floors of clay. This all changed, quite suddenly, when Mars lost its magnetic field. Its ice-fields and glaciers melted, water accumulating behind ice dams. These dams eventually melted too, weakening, collapsing and releasing surges of flood water, which evaporated. Mars became the mainly dry and cold place that it is today.

Given time and better luck in maintaining its magnetosphere, Mars might have developed life, even alien creatures of the sort envisaged by H.G. Wells. If that had happened, we might have been facing a real 'war of the worlds'. But it did not. There is a hope that some kind of primitive life developed early on, in the Noachian era, and survives in niche environments even now. That would be a secret of Mars's life to bring to light!

CHAPTER 7

Martian meteorites: chips off the old block

- Scientific classification: *Phobos and Deimos, satellites of Mars.*
- Distance from Mars: *9,377 km (5,827 miles), 23,460 km (14,580 miles); 0.024, 0.061 times Earth–Moon distance.*
- Orbital period: *7.66 hrs, 30.3 hrs.*
- Diameter: *2 km (1.2 miles), 13 km (8 miles).*
- Rotation period: *Synchronous.*
- Average surface temperature: *−40 °C.*
- Secret debt: *'We were only visiting, but when we dropped in, Mars offered us space and insisted we stay.'*

Mars has two small, irregularly shaped satellites called Phobos and Deimos – Fear and Dread, or Panic and Terror, in Greek. Both satellites were discovered in August 1877 by the US astronomer Asaph Hall at the US Naval Observatory in Washington, DC. Hall had set out deliberately to try to discover whether Mars had moons and had realised that the observational circumstances were particularly favourable that year when the Earth was unusually close to Mars. Mars being so bright, he tried to scrutinise the areas close to the planet, hoping to discern satellites in the glare.

On 11 August, Hall glimpsed a faint point of light near Mars and just had time to measure its position when fog from the River Ptomac rolled in and shut his observing window. Cloudy weather prevented him from working for several days, although he slept at the observatory so as to take advantage of any brief, clear interval. Even when the clouds cleared, a nearby thunderstorm created such bad viewing conditions and the image of Mars was so unsteady that he could not see anything. But he found the satellite again on 16 August. He was so full of his discovery that, in his excitement, he couldn't keep it to himself:

Until this time, I had said nothing to anyone at the Observatory of my search for a satellite of Mars, but on leaving the observatory after these observations of the 16th, at about three o'clock in the morning, I told my assistant, George Anderson, to whom I had shown the object, that I thought I had discovered a satellite of

Mars. I told him also to keep quiet as I did not wish anything said until the matter was beyond doubt. He said nothing, but the thing was too good to keep and I let it out myself. On 17 August between one and two o'clock, while I was reducing my observations, Professor Newcomb came into my room to eat his lunch and I showed him my measures of the faint object near Mars which proved that it was moving with the planet.

Hall discovered the second moon later that night:

For several days the inner moon was a puzzle. It would appear on different sides of the planet on the same night, and at first I thought there were two or three inner moons, since it seemed very improbable to me at that time that a satellite should revolve around its primary in less time [7 hrs 39 mins] than that in which the planet rotates [24 hrs 36 mins]. To settle this point, I watched this moon throughout the nights of 20 and 21 August, and saw, in fact, that there was but one inner moon.

Hall immediately grasped the strange behaviour of the moons as seen from Mars:

The peculiar appearance of these two moons to an inhabitant of Mars is evident upon the slightest consideration. On account of the rapid motion of the inner moon, it will rise in the west and set in the east, and, meeting and passing the outer moon, it will go through all its phases in [seven hours, twice per Martian day].

The names of the moons, Deimos for the outer moon, and Phobos for the inner, were suggested to Hall by Henry Madan, a teacher at Eton College, using Book XV of the *Iliad* as his source:

Mars smote his two sturdy thighs with the flat of his hands, and said in anger, 'Do not blame me, you gods that dwell in heaven, if I go to the ships of the Achaeans and avenge the death of my son, even though it end in my being struck by Jove's lightning and lying in blood and dust among the corpses.' As he spoke he gave orders to yoke his horses Panic and Terror, while he put on his armour.

The Madan family has the distinction of having given names to three worlds, through classical allusions. Henry Madan was the brother of Falconer Madan, the librarian of the Bodleian Library at the University of Oxford. Falconer's eleven-year-old granddaughter, Venetia Burney, suggested the name Pluto (Chapter 16) for that world;

Pluto was the ruler of the underworld and the world is remote from the Sun, so it is cold and dark, which is how the Greeks imagined Hell.

Both Martian moons are small, Phobos being the larger. They are not at all spherical, but have the potato-like appearance of asteroids. Indeed, one theory of their origin is that they were asteroids captured when they passed too close to Mars. Phobos orbits so close to Mars (only about 5,800 kilometres above its surface compared to 400,000 kilometres for our Moon, 3,600 and 250,00 miles respectively) that tidal forces generated by Martian gravity are dragging it down. It is approaching Mars by nearly 2 metres (6 feet) every century. Either it will break up into small pieces and create a system of rings like those of Saturn, or it will crash onto Mars in 50 million years. Phobos's life will end early in one spectacular event or another.

The surface of Deimos is blanketed by pulverised rock and dust, and smooth, except for impact craters. The largest feature on the surface of Phobos is a crater with a diameter of about 9.5 kilometres (6 miles). It is called Stickney, which was the maiden name of Asaph Hall's wife. The surface nearby is covered with about a dozen systems of grooves and streaks. They radiate from the area that leads Phobos in its orbit (its nose, if you think of

Phobos having a face that points in the direction of travel). One theory is that the grooves were formed by boulders rolling away from the site of the impact that created the Stickney crater. Another theory is that the grooves have been formed by a number of collisions between the moon and other rocks that orbit Mars, just as the front of a car would be scratched if driven at speed through grit being thrown out to treat an icy road. These rocks were perhaps ejected into space from the surface of Mars itself.

The solar system is littered with fragments of rock from Mars generated in this way. Over a hundred of them have been discovered on Earth, having fallen as meteorites. This family of small pieces was shot into space from the surface of the Red Planet by the impact of asteroids. 'Chips off the old block' is an indulgent expression dating back at least to the seventeenth century for children who are like their father, an expression that aptly describes these Martian progeny.

The first Martian meteorite seen to have fallen to Earth fell at 8:30 in the morning with a sound like the discharge of numerous muskets on 3 October 1815 near Chassigny, in the Burgundy region of France. It left a smoking trail. A man starting work early in the day in a nearby vineyard saw something fall from the cloud with a hissing sound, like a passing cannonball. (This occurred as France ended decades of the Napoleonic Wars; military noises like muskets and cannon would have been familiar to too many Frenchmen.) The viticulturist ran to see what it was. In a small hole in the freshly ploughed ground, he collected stones, hot to the touch as if warmed in direct sunlight. The stones proved to be meteorites.

A second Martian meteorite was seen and heard to fall on 25 August 1865 by Hanooman Singh near Shergotty in the state of Bihar, India. It was retrieved by W.C. Costley, the Deputy Magistrate of Shergotty, with the aid of T.F. Peppe, the Sub-Deputy Opium Agent for the region. The region was a centre for processing opium grown in the surrounding farmland and shipping it to China. Peppe organised this trade on behalf of the British government. In other words, our knowledge of Mars owes something to the help of a government-sponsored drug dealer.

A third fall occurred in a shower not far from Nakhla near Alexandria, Egypt, on 28 June 1911, in farmland around the village, among the okra, cucumbers and strawberries. The meteorites were collected by William Hume, the director of the Geological Survey of Egypt. One man who claimed to be an eyewitness described how one of the meteorites hit a dog, leaving it like ashes. If true, this much-repeated account would be the first and, so far, only recorded instance of an earthling being killed by a Martian. Sadly, the eyewitness described the event as occurring at a place 30 kilometres (20 miles) away from the actual site of the fall, and happening on the wrong day. The account is the exaggerated product of a lively imagination, and truth has spoiled a good story.

The towns of Shergotty, Nakhla and Chassigny give the class of Martian meteorites their designation as SNC meteorites, or 'Snick' meteorites.

The story of how the SNC meteorites came from Mars to the Earth has been revealed by measuring radioactive elements in the rocks, and their decay products, which, after the rocks have solidified, remain trapped. The rocks had most recently been molten 1,370 million years ago. This is much more recently than most meteorites, which solidified 4,000 million years ago, or more, proving the anomalous origin of the SNC meteorites. Their chemical composition is similar to rocks on the surface of Mars, and one SNC meteorite includes bubbles in its glassy material that contained gas with the exact composition of the Martian atmosphere. This proves that their origin was in a magma field on Mars that solidified after a volcanic eruption 1,370 million years ago.

Most SNC meteorites come from a big piece of Mars ejected by an asteroid impact onto that magma field 200 million years ago. The impact probably ejected a shower of smaller fragments into space as well. This event may have been the one that caused the systems of grooves that radiate from the front face of Phobos, or, if not, an event like that one.

The piece of Mars ejected into space 200 million years ago left the planet and orbited as an asteroid in the solar system. Ten million years ago it was itself broken into smaller pieces by a collision with another asteroid. These small bits showered in all directions, orbiting in space for a further 10 million years, some of them recently falling to Earth.

This story shows that, although the planets of the solar system may be separated by tens of millions of kilometres, they are not completely isolated from one another. They exchange material. Material from the Moon has fallen on Earth as what are termed lunar meteorites. Material from Earth has been thrown to the Moon: one such piece has been retrieved back to Earth by the astronauts of *Apollo 14* as a 2-gram fragment embedded in a football-sized rock catalogued as 14321, and known informally as Big Bertha. Just as there is a two-way traffic between the Earth and the Moon, material from Mars has fallen to Earth and material from Earth must have gone to Mars. When the impact that caused the extinction of non-feathered dinosaurs hit the sea-floor just off-shore from the Yucatán Peninsula, fragments were ejected into space from the 150-kilometre-(100-mile)-diameter crater that it made. Sandstone pieces of the Arizona plateau near

Flagstaff likewise flew into space when the 2-kilometre (1.5-mile) Barringer Meteor Crater was punched into what is now the United States. There are craters all over the world, so meteoroids from virtually every country circulate in space, intermingled and jostling, like diplomats at a United Nations cocktail party.

So, some of the soil in your window box or garden – just a little – is from Mars; the carrots you eat contain a sprinkling from the Red Planet. And, just as the Earth is sprinkled with Martian soil, so Mars is sprinkled with the soil of our own planet. Perhaps that soil contained organisms that had developed in our own benign environment. The hardiest could have survived in space and the hardiest of these, having hitched an interplanetary ride and fallen by chance in hospitable places on the Red Planet, could potentially have survived to colonise Mars. We may excitedly discover life on Mars and find out that it came from Earth.

CHAPTER 8

Ceres: the planet that never grew up

- Scientific classification: *Dwarf planet.*
- Distance from the Sun: *2.77 times the Earth–Sun distance, 414 million km (257 million miles).*
- Orbital period: *4.60 years.*
- Diameter: *0.28 times the Moon, 960 km (596 miles).*
- Rotation period: *0.378 days.*
- Average surface temperature: *–105 °C.*
- Secret grudge: *'I could have been a contender as a planet, but Jupiter held me back.'*

On 1 January 2000, like everybody else in the world, I celebrated the start of a new century and a new millennium, even though I knew this was the wrong day to do it

– a year too early. January 1 was certainly the start of a new year, but the number of noughts in the number of the year gave it an apparently greater significance than it deserved. Properly speaking, the twentieth century of the Common Era ended on 31 December 2000, so the twenty-first started on 1 January 2001.

Likewise, 1 January 1801 was the first day of the nineteenth century, as properly reckoned. The new century was marked by the discovery of a new planet, a coincidence hailed by popular acclaim as auspicious, a cause for optimism. In hindsight we know that Europe was about to embark on more than a decade of the battles, famine, disease and economic instability of the Napoleonic Wars. The Four Horsemen of the Apocalypse trumped the new planet, and the optimism proved forlorn.

Be that as it may, the person who discovered the new planet was Giuseppe Piazzi, a monk of the Theatine order. In the afternoon of 1 January, he had dressed in his warmest, longest coat to go to his observatory in a tower in the Royal Palace of Palermo in Sicily. Palermo was a prosperous city and, if current Italian practice is any guide, on New Year's Day its wealthy families would have attended church, and dined well in the middle of the day, with probably a three-hour lunch. As it grew dark, they would have set out on extensive visits to relatives and friends to talk, play card games and gamble. Not so Piazzi. With his fellow Palermitanos, he would have participated in religious rites in the morning, but his afternoon would have consisted of preparation for his night's work. He would have to stand at the eyepiece-end of a telescope under the open roof of his observatory. His balding head would have been exposed to the cold night air, so he would have put on his special observing hat: it had no brim, and did not get in the way as he put his eye close to the telescope. His observing gloves were warm enough, even though his fingertips protruded though the ends so that he could adjust the brass mechanisms of the telescope. Perhaps through the roof opening he could hear Palermitanos in the streets scurrying through the cold of the clear evening to visit and to take part in festivities. By contrast, Piazzi's self-imposed evening activity was more ascetic. He planned to measure the positions of some stars in the constellation Taurus.

Piazzi used a catalogue by the French astronomer Abbé Nicolas Louis de Lacaille as a source list, remeasuring the Abbé's stars to update their positions. Near one of them, Piazzi spotted an uncatalogued star – one that Lacaille had apparently overlooked. He measured its position, and returned to verify his work on subsequent evenings: the star was not in the same place.

[O]n the evening of the 1st of January of the current year, together with several other stars, I sought for the 87th of the Catalogue of the Zodiacal stars of Mr la Caille. I then found it was preceded by another, which, according to my custom, I observed likewise, as it did not impede the principal observation. The light was a little faint, and of the colour of Jupiter, but similar to many others which generally are reckoned of the eighth magnitude. Therefore I had no doubt of its being any other than a fixed star.

In the evening of the 2d I repeated my observations, and having found that it did not correspond [in position] with the former observation, I began to entertain some doubts of its accuracy. I conceived afterwards a great suspicion that it might be a new star. The evening of the third, my suspicion was converted into certainty, being assured it was not a fixed star.

The change of position indicated that the 'star' must be a planet, or possibly a comet, though Piazzi could see no fuzziness or tail. Piazzi calculated its orbit, which approximated to a circle lying between the orbits of Mars and Jupiter, filling in a wide gap in the solar system. Piazzi concluded that he had discovered a new planet. He named it Ceres, after the goddess of agriculture and the protector goddess of Sicily.

It was surprising that the new planet was so faint. Other planets, sometimes considerably more distant, are much brighter. It had to be significantly smaller, intercepting and reflecting much less sunlight.

Over the next year, Ceres was followed intensively in order to improve on its orbit. Unless this is done immediately after discovery, so that a planet's future position can be predicted accurately, it might move on so much when, for some reason, it cannot be seen (for example it moves behind the Sun), that it is lost in the confusion of other stars. If you know fairly accurately where it will be when it reappears you can more easily find it again. At the end of March 1802, Wilhelm Olbers, a prominent doctor in Bremen in Germany, but also a keen amateur astronomer, was making repeated observations of Ceres and the stars nearby when he saw a star that had not been there earlier, in January. He measured the star's position and followed it for two hours. Like Ceres, it also moved.

What was it that Olbers had found? 'What shall I think of this new star?' he wrote. 'Is it a strange comet or a new planet? I do not dare judge it yet. It is certain that it does not resemble a comet in the telescope; no trace of nebulosity or atmosphere around it can be seen.' The new object was named Pallas. It proved to have a similar orbit to Ceres as well as a similar brightness: there were two new planets. True, they were quite small compared to the other planets, and what they were deserved a new classification name. The British astronomer William Herschel, in 1804, suggested 'asteroid'. Two similar planets of a new kind in the same orbit – this called for a special explanation. Olbers came up with one: maybe there used to be one planet that had split into two asteroids. This immediately raised the possibility that there were more than two asteroids; perhaps the original planet had split into three pieces – or four! Or more!

It soon became clear that there were more than two pieces. The German astronomer Karl Ludwig Harding earned his living as a tutor but was an ardent amateur astronomer obsessed with the idea of discovering a new planet. On 1 September 1804, his perseverance paid off; he found a new star in the region where Olbers had predicted any further planets would orbit. It was a third asteroid, which became known as Juno.

Olbers pursued his hypothesis with enthusiasm, and concentrated his search on the region where all the orbits of the three asteroids intersected, presumably the place where the disruption had occurred. On 29 March 1807, in the same area, Olbers found a fourth asteroid, his second, which became known as Vesta. But there were more asteroids to come.

Olbers' hunch had paid off, but, it turned out, was based on an idea that was credible but untrue. Another German astronomer, Johann Huth, suggested in 1804 what has become the main theory in modern times about the origin of the asteroids:

> I hope that this [asteroid] is not the last one that will be found between Mars and Jupiter. I think it very probable that these little planets are as old as the others and that the planetary mass in the space between Mars and Jupiter has coagulated in many little spheres, almost all of the same dimensions, at the same time in which happened the separation of the celestial fluid and the coagulation of the other planets.

Huth's conjecture was remarkably close to the modern idea. Asteroids began life in the solar nebula as small solid bits that stuck together when by chance they collided. They grew to a certain size, at which they became 'planetesimals', massive enough to attract one another and other small lumps nearby, a process called accretion. But Jupiter's gravity had a big influence on the planetesimals orbiting near to it. It stirred up the nebular material so that if a planetesimal drew material towards itself, Jupiter gave the stream an impetus that meant the material flowed past instead of accreting. Ceres was the largest such planetesimal that was able to grow under these circumstances.

Because Ceres is large as asteroids go, it has a strong enough force of gravity to be able to pull the asteroid into a nearly spherical shape. It started off as a rubble-pile – a loose agglomeration of individual planetesimals. But the material gradually settled and consolidated. Two effects contributed to the process of making Ceres spherical. In the first process, when Ceres collided with a meteor, a rock perched on a hill could be vibrated off the peak and would roll down the slope and drop into the dip at the bottom. Gradually the larger hills were ironed out. In the second process, the core of Ceres was large enough that it generated a considerable amount of heat from the decay of radioactive elements inside. The heat was kept inside Ceres's outer layers, which were thick enough to act like a blanket. The temperature of Ceres rose and partially melted rock in the body of the planet. The heavier material, such as minerals with a high iron content, sank down; the lighter material floated up. As a result, Ceres has 'differentiated' into a sphere of layers with different mineral characteristics. There is a rocky inner core, rich in metallic ores, and an icy outer mantle. Ceres is what a 'rubble-pile' spontaneously turns into if the pile grows large enough.

If Jupiter had allowed Ceres to develop by accreting more planetesimals around it, it would have become an

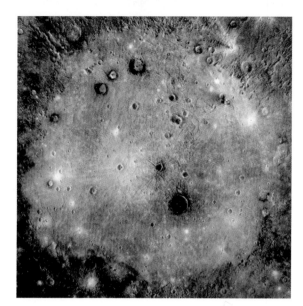

ABOVE: 1. The Caloris Basin on Mercury, imaged by the *Messenger* space probe, is an old crater filled with lava that rippled out and swamped old craters and with more recent smaller craters overlying its floor. Colour in the picture has been manipulated to differentiate minerals (lava is tan-coloured, for example). BELOW: 2. Lava flows on Venus extend for hundreds of kilometres from the base of the 8,000-metre-high volcano Maat Mons across the fractured plains in the foreground. In this radar picture by the *Magellan* spacecraft, the terrain is accurate, but its vertical scale and colour are exaggerated, and in reality the sky is not black.

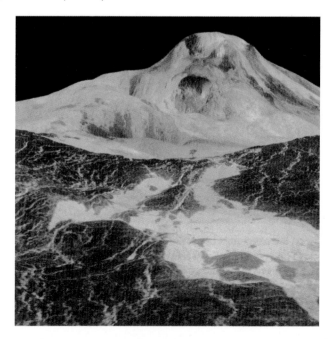

3. On their way to the Moon the astronauts of *Apollo 17* snapped this picture with a handheld camera, looking back to the Earth. It shows a remarkably cloud-free Africa, with an ice- and cloud-covered Antarctica. India is passing into night at the right-hand edge.

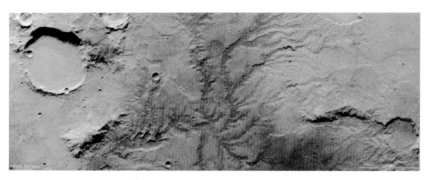

4. The *Viking Orbiter* spacecraft imaged networks of dry valleys on Mars, which show that at one time rivers flowed over what is now a desert.

5. As the warmth of the day loosens the grip of ice on sand-dunes near the north pole of Mars, sand slips down the slopes and exposes darker material underneath the surface. The photo caught the billowing rosy-red dust from a landslide that had just happened (below and left of centre).

6. The large crater Stickney is the most prominent feature on Phobos, the larger moon of Mars. Scratches radiate from it, perhaps from boulders rolling downhill, or caused when Phobos flew through a cloud of rocks ejected from Mars. Overall, Phobos has the potato-like shape of an asteroid. Imaged by the *Mars Reconnaissance Orbiter.*

ABOVE: 7. Imaged by the *Dawn* spacecraft, the side of a crater on Ceres has collapsed in a landslide as the ice binding the soil of this asteroid loosened its hold. BELOW: 8. Clouds stream by Jupiter's Great Red Spot in this picture from the *Juno* space mission.

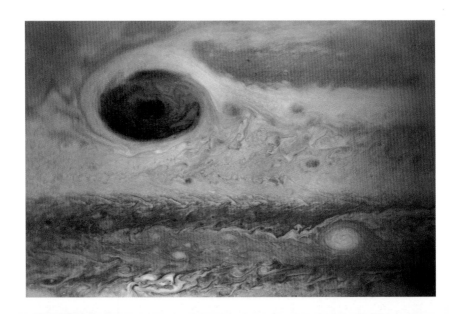

ABOVE: 9. When the *New Horizons* spacecraft flew by Jupiter in 2007, it caught an eruption on Io, its 300-kilometre-high ash cloud spewing from the volcano Tvashtar. BELOW: 10. The *Galileo* spacecraft imaged Europa and showed its icy plains, cracked and grooved, the gaps stained by evaporating salts and its ice floating on a salty lake underneath.

ABOVE: 11. Saturn's icy rings shine in scattered sunlight in this view by the *Cassini* space probe, which looks toward the unilluminated northern side of the rings. BELOW: 12. Pan formed within Saturn's rings, accreting ring material and forming the rounded shape of its icy, central mass when the ring system was younger and thicker. It accreted the thin ridge around its equator when material was raining down more recently, and the rings were thinner.

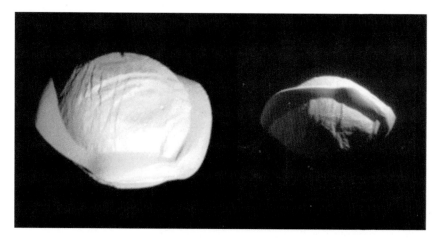

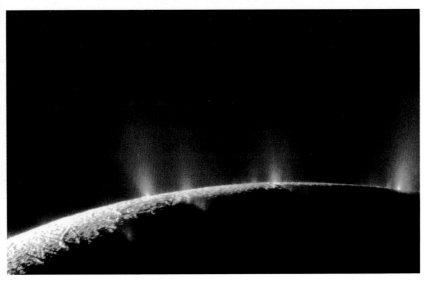

ABOVE: 13. Clouds of icy chips are backlit by the Sun, showing their eruption in geysers from the interior of Enceladus out through many locations along the Tiger Stripes. BELOW: 14. The *Huygens* lander took one picture on the surface of Titan across the lake bed on which it landed: boulders of ice littered the lake floor below a brown smoggy atmosphere.

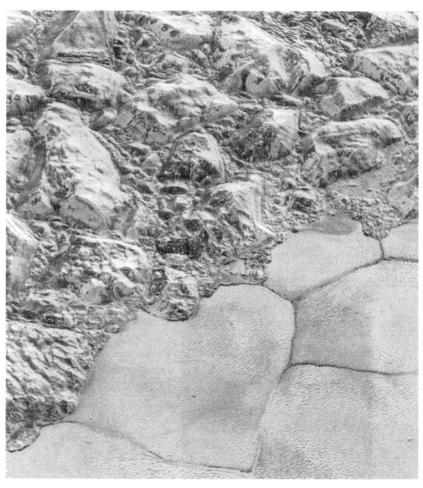

15. Imaged by the *New Horizons* spacecraft, great blocks of water-ice jostle together in a mountain range on Pluto up to 2 kilometres high. The mountains end abruptly at the shoreline of Sputnik Planum, a plain within a large crater on which the textured surface of soft nitrogen ice forms a nearly level surface, broken only by striking, cellular boundaries.

Earth-like planet. The strong gravitational pull of Jupiter, however, stirred up the planetesimals and inhibited this process. Ceres stopped eating, starved by the malign influence of Jupiter. It succeeded in becoming an immature planet but could not make the last steps to get larger, mature and thus dominate all the other smaller bodies near its orbit.

Although Ceres is a case of arrested development, it did grow up enough to become a so-called 'dwarf planet'. This is a planet that, although it has settled into a spherical shape, has missed out on full planet status because it has not eaten up everything in the same orbit. Like Peter Pan, Ceres never grew up.

The number of asteroids now known is huge. There may be as many as 2 million asteroids larger than 1 kilometre (0.6 miles) in size and 25 million asteroids larger than 100 metres (100 yards). About three-quarters of a million are known and catalogued. Why so many small asteroids? The answer is: 'because there were so many large asteroids to start with'. There were so many asteroids confined into a restricted area of the solar system that collisions were inevitable. The fragments from the collisions between large asteroids became the numerous smaller asteroids.

Asteroids nowadays are thus a mixture. Some are primitive planetesimals, bits of material from the original solar nebula that never grew; one could say, stillborn planets. Some, like Ceres, are planets in arrested development. Some were comets that have visited and revisited the Sun so often that all the ice they contain has blown off as gas and is now exhausted, leaving behind only rocky material: dead comets. Some, perhaps most, are fragments of larger asteroids that collided and broke – injured veterans of past combat.

The larger asteroids have been well studied by space probes. There has been a considerable number of fly-bys by space probes *en route* to other destinations, but the first spacecraft dedicated to an asteroid was NASA's *NEAR Shoemaker*, which entered into orbit around Eros on Valentine's Day 2000, landing on its surface in 2001. The Japanese Aerospace Exploration Agency's *Hayabusa* probe studied Itokawa in 2005. A second JAXA probe, *Hayabusa2*, went to Ryugu in 2018. NASA's *Dawn* spacecraft orbited Vesta in 2011–12, and went on to Ceres in 2015. NASA's *OSIRIS-Rex* was launched in 2016 and is visiting Bennu. If all goes well, it will return a sample from its surface to Earth in 2023.

Although small for a planet, Ceres is the largest asteroid by a considerable margin. It is an icy, rocky planet some 950 kilometres (590 miles) in diameter, rotating with a 'day' of nine hours. Pictures from the *Dawn* space probe show a world that is like our Moon. It has a large number of craters made by meteor impacts, and a number of bright spots, some of which become hazier from time to time. The occasional haziness suggests that Ceres is still geologically active, venting gases and ash or dust.

The most surprising finding by the *Dawn* probe is that some of the bright spots are white, salty deposits, made mainly of sodium carbonate that made its way to the surface in a slushy brine from within or below the crust, the traces of an ancient ocean. The data suggest there still may be liquid under Ceres's surface and that some regions are being fed from a deep reservoir. In the region of the crater named Ernutet, organic molecules were found in abundance. Organic molecules are ones that contain carbon. They are the sorts of molecules made by life, although they do not have to have been made in this way. The carbon-rich compounds are mixed with minerals that are the products of the interaction between rocks and water, such as clays. These deposits have oozed from Ceres's interior, having been made some time ago in the early, interior ocean.

Vesta has a diameter of 530 kilometres (330 miles), so it is considerably smaller than Ceres. It is much less spherical than the latter – its surface gravity is not strong enough to make Vesta quite qualify as a dwarf planet. The Hubble Space Telescope was able to see that Vesta has a gigantic piece missing at its south pole, two giant, overlapping craters, something that was confirmed when *Dawn* arrived for a close inspection. One of the craters is relatively recent.

Vesta is the largest, brightest member of a family of much smaller asteroids, 10 kilometres (6 miles) in size, that have identical orbits. Using the same argument that had earlier been proposed by Olbers in relation to Ceres and Pallas, but which had proved not to be so, this suggests that the Vesta family really are related, the result of a catastrophic incident that broke a single body into many fragments. There is supporting evidence. Some of the smaller pieces from this collision fall to Earth from time to time as meteorites of a kind called HED meteorites, which are linked to Vesta because the composition of the meteorites matches the asteroid's surface composition as measured by the *Dawn* spacecraft. HED stands for the distinctive meteoritic minerals howardite, eucrite and diogenite, and these exist also on Vesta.

The natural inference is that the HED meteorites and smaller asteroids originated in the impact of a big meteor on Vesta, which scattered fragments of Vesta everywhere and excavated the recent giant crater.

Ceres has been lucky. Over the last 4 billion years, it has collided with some of its smaller neighbours, but no big ones. Vesta has been lucky too, in a less lucky way. It did have such a collision or two, but survived. Other asteroids have not been so fortunate. They suffered collisions that were so powerful that the colliding asteroids were fragmented into bits. The bits are much too small to settle into a spherical shape. They orbit now as jagged, angular, solid shapes. The fragments are made of iron and other heavy metals if they came from the central cores of the pre-collision asteroids, or of stone if they came from the outer mantles.

The bits from the two kinds of fragmented asteroids sometimes fall to Earth as different kinds of meteorites. There are two main kinds: iron meteorites and stony meteorites. Iron meteorites come from the differentiated core of a broken asteroid. They are dense and surprisingly heavy for their size. Stony meteorites are the most common type, and were once part of the outer crust of a differentiated asteroid. They look much like any other stony rock. Iron meteorites that have fallen to Earth over the millennia have a black surface and are easy to spot, especially in a rocky sandy desert.

Meteorites are sought after by collectors. When cut and polished to display their interior minerals, they are often beautiful. They have a technical appeal because of the various minerals of which they are made, and there are some rare types that it can be satisfying to own because other people do not. They also have a romantic appeal: it is mesmerising to hold a meteorite in your palm and visualise its birth and its long history voyaging around the solar system. As a result of these meteorites' appeal, there are dealers who seek them out for resale. They hunt for them, converging on places where a large meteorite has been reported in the hope of finding pieces that have broken off. They cover large areas of desert looking for meteorites that have been seen to fall. It can be a rewarding treasure hunt: the most expensive meteorites sell for sums upwards of $500,000.

Meteorite hunters can collect iron meteorites by scanning a desert plain, like the Nullabor in Australia, or the Karoo in South Africa. One dealer covers large areas by sailing in a hang glider at a low elevation. Stony meteorites are much more difficult to find in such places because they merge in with everything else. However, both kinds of meteorites are readily picked out on snow, which is why the Antarctic continent has become the favourite venue for meteor hunters.

The provenance of a meteorite adds another dimension to its potential appeal to collectors – a massive cosmic collision, a billion years in lonely space, a fiery descent to Earth and a millennium in snowy wastes, perhaps. When I gaze at a meteorite held in my palm, I imagine a life that, considered against cosmic time-scales, is as brief as a firefly's: fleeting, but grander and much more spectacular.

CHAPTER 9
Jupiter: hard hearted

- 🌐 Scientific classification: *Gas giant planet.*
- 🌐 Distance from the Sun: *5.20 times the Earth–Sun distance, 778.6 million km (483.8 million miles).*
- 🌐 Orbital period: *11.9 years.*
- 🌐 Diameter: *11.21 times Earth, 142,984 km (88,846 miles).*
- 🌐 Rotation period: *9 hrs 55 mins.*
- 🌐 Average temperature of the top of the clouds: *−110 °C.*
- 🌐 Secret complaint: *'I am the supreme ruler of the solar system but I never get any peace – I've had a headache from that storm for the last 450 years, and those pesky comets keep jabbing me.'*

Jupiter is named after the ruler of the gods of classical Rome, the same god as the Greek Zeus. It is the most massive of the planets, the most important. The ancients who made the connection between the god and the planet could not have known its size and must have inferred its status from its brightness (it is, at its brightest, the second-brightest planet after Venus) and its stately motion.

Although it is much less massive, Jupiter's gravitational pull on the Sun is so strong that the Sun is not really stationary in the middle of the solar system: instead, the two objects revolve around their centre of gravity, a point which lies near the surface of the Sun. If there are in our galaxy alien civilisations with astronomers, they could find out about our planetary system, at least about Jupiter, by measuring the oscillating motion of the Sun, with its twelve-year period linked to the orbital revolution of Jupiter.

Jupiter is covered with clouds. Because it is bright and its image is large, and the clouds alter rapidly as the weather and the climate on Jupiter change from hour to hour and from year to year, it is rewarding to view the cloud tops even through an affordably sized telescope. Lots of amateur astronomers do this, following what they can readily see. At the other extreme of visibility, entirely hidden at Jupiter's centre, there is an exotic substance. It is obscured literally, and also hidden metaphorically, at the boundaries of scientific knowledge. Jupiter and Saturn are the only places in the Universe where this substance is known to exist – at least, where there is good evidence that it exists. A few scientists in a small fraternity continue to attempt to uncover the details of this substance.

Jupiter was born and now lives beyond the snow line of the solar system. The solar system had been created from a slowly rotating cloud of gas, ices and solid particles of dust. The new-born Sun warmed and made gaseous the ices in the closest parts of the cloud. Its warming power reached only so far, and the ices persisted in the outer regions of the solar system. 'Outer regions' means the regions beyond the snow line.

The term 'snow line' is borrowed from geography, where it means the contour line on a mountain above which it is perpetually so cold that snow never melts. In astronomy, it means the orbit in a planetary system beyond which icy material is not melted by the parent star. Further out from this orbit in our solar system, the huge planets, Jupiter, Saturn, Uranus and Neptune, were born 4.6 billion years ago, drawing in unvaporised ices but also the lightest gases, hydrogen and helium. These gases had completed their journey from the first minutes of the Universe 13.6 billion years ago to these planets, a journey 9 billion years in the travelling.

The two lightest gases are by far the most abundant materials in the Universe. Moreover, at the distances from the Sun that lie around the snow line, there was in the solar nebula a lot of material. When this material ends up in a planet, it becomes massive. This happened to Jupiter, Saturn, Uranus and Neptune. The scientific description for these planets is 'gas giant', a term that is self-explanatory.

Jupiter is primarily hydrogen and helium. It has no solid surface as such, which is what makes the term 'terrestrial planet' inappropriate. Jupiter is huge – ten times the size of the Earth, and 300 times as massive. It rotates very fast – once every nine hours and fifty-six minutes: the quickest rotation of any planet. Because of this, Jupiter has a distinct belly, like a dissolute monarch: the planet visibly bulges at the equator, and is flattened at the poles. Its equatorial radius is 4,600 kilometres (2,900 miles) more than its polar radius.

If Jupiter were just 30 per cent more massive, it would be a star, albeit a star of a rather feeble kind known as a brown dwarf. The definition of a star is that it generates heat from nuclear reactions in its hot, dense interior. If Jupiter were an ordinary star, it would do this via nuclear reactions that burn hydrogen. If it were a brown dwarf, it would burn helium. Jupiter does neither, so it is not a star.

Jupiter may rule the solar system of planets, but its power is limited. If it is a ruler, it lives the life of a warlord: there are greater powers all around the Galaxy.

Going inwards from the top of the clouds, Jupiter's hydrogen and helium atmosphere gets denser and denser and turns into liquid. At the very centre of Jupiter is a dense, presumably rocky core perhaps ten to fifty times the mass of the Earth. Between the two is a progressively denser zone of hydrogen and helium and other gases, probably mixed with rock and ice in a kind of slushy, icy mixture that gets progressively thicker towards Jupiter's centre.

At the top of Jupiter's atmosphere are multicoloured clouds, arranged in zones of alternating light and dark hues that run along the lines of latitude. The bands are alternately upwelling and down-falling atmospheric gases. The clouds are coloured in reds and yellows by droplets and particles of strange chemicals, the origins of which are controversial. In general, it seems that the lighter clouds are higher than the darker ones, so the colours must be brought up from within Jupiter.

The colour occurs as if blood has rushed up to the skin of Jupiter's face in a blush, rather than being created by the chemical action of sunlight on the top of the clouds, as if Jupiter is getting a tan.

A cosmic accident brought some of the chemicals into view in 1994. Comet Shoemaker-Levy 9 passed too close to Jupiter and broke into more than twenty fragments. The fragments shot past Jupiter but were pulled back and, two years later, plunged one by one into Jupiter's atmosphere. Each fragment tunnelled into the atmosphere so quickly that a hollow tube was temporarily formed in their place. Gases from below welled up each tube and sprayed out of it, like water from a fountain. The spray of gases flowed in an arc onto the cloud tops. Dark-coloured chemicals had been brought up from the cloud layers below, such as sulphur, carbon disulphide, ammonia and hydrogen sulphide. The dark-coloured stains on the lighter cloud tops persisted for months.

Hydrogen sulphide is notorious for its smell of rotten eggs: schoolchildren use it for stink bombs. The other chemicals – combinations of sulphur, nitrogen and hydrogen – also have a strong odour. In other words, Jupiter smells.

The largest feature of Jupiter's clouds is called the Great Red Spot. It is elliptical in shape, 24,000–40,000 kilometres (15,000–25,000 miles) east-to-west and 12,000–14,000 kilometres (7,500–8,500 miles) north-to-south. The Earth could comfortably fit inside the elliptical shape. It is a gigantic storm, towering much higher than the cloud tops around it, a high-pressure anti-cyclone. It was first seen by Giovanni Cassini in 1665, a 'permanent spot' that was followed by astronomers from its discovery to 1713. There is a gap of a century in the record of anyone seeing it until 1830; it is not known if it disappeared or whether it remained there, perhaps faint, and everyone missed it. Whatever the truth about that gap in its history, the Great Red Spot has been there ever since – it is the mother of all storms and has probably lasted 350 years.

According to classical mythology the god Jupiter hurled thunderbolts at anyone who displeased him. The planet generates lightning too, as discovered by the *Voyager 1* and 2 spacecraft as they flew by Jupiter in 1979. Looking back at the dark, night side of Jupiter, they saw flashes of lightning in massive thunderstorms, illuminating the clouds. The *Galileo* spacecraft confirmed their discovery in 1997. As on Earth, the lightning is created by moist clouds rubbing together in the water layer of Jupiter's atmosphere, about 100 kilometres (60 miles) below the cloud tops.

Jupiter has a strong magnetic field, fourteen times stronger at the cloud tops than the Earth's. Like Earth's magnetic field, Jupiter's originates in circulatory motions in its interior. However, Jupiter is not made of iron, but principally hydrogen, so how come? There is uncertainty about the internal structure of Jupiter but the 'best buy' theory at the moment is that the interior of the planet below about 20,000 kilometres (12,000 miles) all the way to the rocky core is made of 'metallic hydrogen'. This exotic form of hydrogen was theoretically predicted in

1935 by US physicists Eugene Wigner, later a Nobel Prize-winner, and Hillard Bell Huntington, then a student. It is formed when hydrogen gas is compressed at extremely high pressures. The gas molecules are forced to arrange themselves into a lattice, somewhat like a crystal, that acts like a metal and conducts electricity, just as iron or copper do.

Making metallic hydrogen in a laboratory so that it can be experimentally examined is regarded as one of the holy grails of physics. There have been several claims that small samples have been made, but not everyone has been convinced that the claims are proven. The pressure that is needed is at the limit of what can be achieved on Earth, perhaps even higher. Metallic hydrogen remains a substance that, so far, can be studied only in theory. The gas giants Jupiter and Saturn are the only known places in the Universe where scientists can probe indirectly, through more theory, the secrets of metallic hydrogen's properties.

It is not ideal when the subject of scientific study is out of reach, below 70,000 kilometres of a planet that lies 600 million kilometres away. But the Universe provides laboratories that are more extreme than terrestrial limitations, often at the edge of our imagination, by making exotic situations that scientists can use to probe otherwise unattainable conditions. The key condition in this case is high pressure. And 70,000 kilometres of a planet has to be supported by a high internal pressure – in the region of a million times the atmospheric pressure on Earth!

The magnetic field generated in the metallic hydrogen forms Jupiter's 'magnetosphere'. It acts like a bottle, both keeping out electrical particles that originate in the Sun, and keeping in particles that originate from the planet. Particles zoom about inside, bouncing off the walls of the magnetospheric bottle, and generate radio waves – Jupiter was one of the first celestial objects identified as a radio source by the pioneer radio astronomers in 1955. In addition, Jupiter has strong displays of aurorae, caused by electrical particles zooming down the magnetic field lines and crashing into the atmosphere near the poles. Solar particles cannot push into the magnetospheric bottle; they cannot get through the stronger magnetic field that surrounds the planet, close to it. However, the magnetic field does extend out into space by threading up around the poles. There are routes at the poles along which electrical particles can crash down onto the atmosphere, channelled along the polar magnetic field. As on the Earth, the solar particles are guided to impact on the atmosphere in a circular shape called the 'auroral oval', the zone of greatest auroral activity.

Earth's auroral oval varies in size but is about 10 to 20 degrees of latitude in radius, about 2,000 kilometres. It is centred on the magnetic pole. In the north, the magnetic pole lies at the present time over the Arctic Ocean, off-shore from Ellesmere Island in the very north of Canada. The auroral oval itself typically runs over northern Norway, the southern tip of Greenland, along the Canada–USA border, through Alaska and along the Arctic coast of Russia. If you are thinking about a trip to see the aurora, these are the best locations, on average. (On the World Wide Web, there are so-called space-weather services that attempt to help auroral tourism by making forecasts of auroral activity and location if you want to fine-tune your choice of viewing station.) Jupiter's auroral oval is the same size as expressed in degrees of latitude on Jupiter's cloud tops, but ten times larger as measured in kilometres, so the oval is huge, as large as the entire Earth.

A unique feature of Jupiter's auroral oval is that it contains features that are from Jupiter's four large moons: Io, Europa, Ganymede and Callisto. As the moons orbit over Jupiter, they interact with its magnetic field. Each moon is surrounded by a kind of atmosphere, material ejected into space, such as Io's volcanic emissions. Each moon thus feeds electrically charged particles into Jupiter's magnetosphere. The particles run directly down the magnetic field onto Jupiter's cloud tops and make spots of aurorae where they impact. The spots rotate around the pole, footprints below the satellites as they revolve in orbit.

Io is the principal source of the material that feeds Jupiter's magnetosphere. Radio emission from Jupiter comes in bursts, depending on how actively Io is ejecting material. But the strength of the radio waves depends also on Io's position around Jupiter. Changes of the position cause corresponding changes of the intensity of the radio emission, which therefore varies with the same period as Io.

The footprint of Io in Jupiter's auroral oval is the brightest of the four. All the other three satellites have footprints too, but weaker. Callisto is the most distant moon, its auroral footprint is the weakest, and the place where it occurs is confusingly superimposed on brighter parts of the auroral oval. It is not easy to see it and it was identified in 2018 only after an intensive search of archive images captured by the Hubble Space Telescope.

It is hard to see Jupiter's poles from Earth, and, although the Hubble Space Telescope is in space, it is in an orbit not far above the Earth's surface, so its viewpoint of the auroral oval is not the best. The *Juno* spacecraft, which entered orbit around and over Jupiter in 2017, has an instrument on board that is specifically designed to study Jupiter's aurora. It has seen that Ganymede's auroral footprint is double, two spots separated by 100 kilometres.

Ganymede is the only moon of the four that has its own magnetic field, and the double structure is something to do with the shape of Ganymede's magnetosphere.

Our Jupiter serves to compare and contrast with similar planets that have been found in other planetary systems. About a thousand planets as large as Jupiter are known. Around half of them are at Jupiter-like distances from their parent star, but the other half are 'hot jupiters', much closer, and much warmer. They are so hot that they are evaporating. They cannot have formed where they are now, which would have been beyond the snow line. Somehow, they have migrated inwards, fleeing the cold and apparently seeking the warmth. In this, the extrasolar jupiters serve as models for our own Jupiter and provide an insight into some secrets of its early life.

There are two interactions in early planetary systems that might have been responsible for moving jupiters and making them hot. The earliest is an interaction between the jupiter and the disc of gas and dust left over from the formation of the planets. As the planet grows, it can open up a gap in the disc or create concentrations in the disc. Any asymmetries that are thus created pull the planet off course and it can migrate. In the cases of some hot jupiters, they migrate a long way inwards and get as close to their sun as we are to ours, or even closer. In our solar system, Jupiter started to do the same but its journey stopped early.

The second interaction that causes planets to migrate occurs later, at the time when the jupiters have fully formed, in company with a large number of planetesimals. The planetesimals move everywhere among the giant planets. The jupiter may encounter some of them closely and eject them from the planetary system. As they are ejected, they give the jupiter a little backward kick, and it gradually migrates inwards, towards its sun. According to the Nice Simulation (Chapter 2), this happened in our solar system to both Jupiter and Saturn.

Jupiter is the ruler of the planets, but not a ruler that has total power over his own destiny. Jupiter is the largest single influence on the dynamics of the solar system after the Sun, but the solar system reacts back on Jupiter. In a royal court, the monarch has more power than any other member, but, except for the most absolute of rulers, the monarch is subject to forces arising from the action of the courtiers. The planets have individual personalities and lives, but together they make up a planetary system that acts as a community in which some members, like Jupiter, have greater influence than others.

CHAPTER 10

The Galilean satellites: siblings of fire, water, ice and stone

- Scientific classification: *four main satellites of Jupiter.*
- Distance from Jupiter: *422,000 km (262,000 miles) to 1,880,000 km (1,168,000 miles); 1.09 to 4.90 times the Earth–Moon distance.*
- Orbital periods: *1.77 to 16.7 days.*
- Diameters: *3,650 km (2,270 miles) to 4,820 km (3,000 miles); 0.286 to 0.378 times Earth.*
- Rotation periods: *Synchronous.*
- Average temperatures: *−155 °C.*
- Secret grievance: *'The planets receive more attention than us satellites, but we have as much variety and we are more hospitable than most planets.'*

Even with just a pair of binoculars you can see the four main satellites of Jupiter. They are not planets, because they orbit their parent planet rather than the Sun, but they are certainly planet-like. They are all roughly 3,000 to 5,000 kilometres in diameter, the smallest being slightly smaller than the Moon, the largest slightly larger than Mercury, and they are all similar to the terrestrial planets.

Three sisters and a brother, they form a motley crew. As siblings, the satellites have a family resemblance but are all different. Together with Jupiter they make a miniature planetary system, and, it is thought, were formed at the same time as Jupiter in much the same way that our solar system was formed. Fundamentally, they are rocky, but their position far out in the solar system means that, with one exception, they have retained the original ices they accrued from the solar nebula. Water and ice have played a considerable part in their lives, although one of them now lives a dry life of fiery eruptions.

We on the Earth lie in the same orbital plane in which the four satellites revolve around Jupiter, so to us they seem to line up in a row, moving back and forth from side to side, sometimes passing in front of Jupiter and casting a shadow on its cloud tops and sometimes hiding, either obscured behind it or in its shadow. Their periods lie roughly between a day and two weeks, so they change their positions while you watch them from night to night, or even from hour to hour through the night. When they move into Jupiter's shadow and are eclipsed, their light is extinguished in a few minutes, fading progressively until what is called the 'last speck' of light is extinguished.

The satellites remain, as seen from Earth, as points of light. But we know their structure from the visits of four space probes. The *Voyager* spacecraft, *1* and *2*, flew by Jupiter in 1979. *Galileo* was the first spacecraft to enter into orbit around Jupiter in 1995 and was able to make extensive observations over eight years. A second probe, *Juno*, entered orbit around Jupiter in 2016. These space probes have revealed the landscapes of Jupiter's satellites as, in one case, a volcanic desert, and, in the others, an Antarctic continent of rocks, icebergs and cold, cold oceans.

The four are called the Galilean satellites, because they were discovered by Galileo in the first two weeks of 1610, with his new telescope. On the first night he saw only three stars, two on one side of Jupiter and one on the other. He saw three again on the second night, but all of them were on the same side of Jupiter. He thought at first that they were a chance line of three stars and that the change was due to the motion of Jupiter through the three. A couple of nights later there were only two and then a few nights later there were four.

Initially Galileo thought that the four moved back and forth in a straight line. How could they pass through the body of Jupiter? But suddenly the penny dropped and Galileo realised that the four 'stars' were moons in orbit around Jupiter. It was a dramatic discovery because it disproved the theory that every celestial body orbited the Sun. In fact, the way the satellites orbited Jupiter was an exemplar for the way the planets orbited the Sun, as in the theory formulated by Copernicus in 1543.

Galileo referred to the four simply as I, II, III and IV but christened the satellites as a group. He called them 'the Medicean stars', in the hope that Cosimo II de' Medici, the seventeenth-century Grand Duke of Tuscany, would become his patron. His plan worked – Cosimo appointed Galileo as his Philosopher and Mathematician and provided a stipend. But Galileo's collective name for the satellites was rejected by other astronomers, who did not look well on naming stars after someone else's patron. They became known as Io, Europa, Ganymede and Callisto, all of them in mythology Jupiter's lovers (of both sexes).

The closest of Jupiter's moons to the planet is Io. The entire surface of Io is coated with black rocks mixed with sulphur – yellow, orange and red in its different forms – like a medieval picture of Hell. There are almost no meteor craters, showing that the surface is young, and that geological processes have erased craters that had formed earlier. However, the surface of Io is pitted, like an acne-scarred face. The pits are mostly not meteor craters. They are calderas, among mountains taller than any on Earth, and lava flows, some of them cold and solid, others hot and flowing. It is a volcanic landscape. I imagine it to be like the volcanic landscapes of La Palma in the Canary Islands and the Big Island in Hawaii, the observatories of which I worked in for several years. The ground is made up of solidified mounds of smooth, black lava and loose, jagged rocks. Where the land has been sectioned by the explosion of a volcano that had excavated a crater, drifts of yellow or orange ash are exposed. In the still-active areas, there are vents of steam and sulphurous gas, heat radiating from the red, molten lava oozing from below.

As the closest of the satellites to its parent planet, Io is squeezed and relaxed by strong tidal forces originating in Jupiter. The heat generated by this repeated working of Io's interior has melted its rocks and created about four hundred volcanoes, some so active they shoot lava 400 kilometres (250 miles) into the sky.

The volcanoes on Io were discovered by Linda Morabito, the navigation engineer for the *Voyager 1* space probe. During its flight through Jupiter's system of satellites, Morabito's task was to identify stars in images taken by

177

the navigation camera, determine the position of the spacecraft and correct the spacecraft trajectory in real time so that it did not crash into something. Later, the images would be analysed to reconstruct the trajectory even more accurately, as a basis for stitching together images of the planetary surfaces. As the encounter was ending and the spacecraft was tracking away from Jupiter and its satellites, there was behind them a particular star that was key to the accuracy of the navigation process. The star was dim and Morabito had to process its images by 'stretching' them, increasing their contrast so that she could see it. She noticed something that was invisible on the unprocessed picture. It was a kind of cloud, very large, just above Io's surface. The 'cloud' was positioned over a heart-shaped feature on Io.

What Morabito had discovered was an ash-cloud from a volcano, now named Pele (after the Hawaiian goddess of volcanoes), and the heart-shaped feature was the volcano itself, with its slopes, ejecta and lava flows. 'I had the sense that I was seeing something that no one else had seen before,' she recalled. Later that day at dinner she had the satisfaction of telling her parents that she had discovered the first volcanic activity outside the Earth that anyone had ever witnessed.

Io is only a little larger than our own Moon. It is slightly ellipsoidal (the shape of a rugby football, or an American football); Jupiter's tidal forces have locked on to the long axis, which points towards the parent planet, so that, like all the Galilean satellites, Io gazes at Jupiter with the same face all the time. Its rocky surface is all but ice free (presumably the volcanic heat has evaporated all the water) and is coated with sulphur: its colours are the colours of different forms of sulphur. Material ejected by the volcanoes forms a thin atmosphere, and feeds into Jupiter's magnetosphere. The volcanoes generate lava flows hundreds of kilometres long and hundreds of times the volumes of recent flows from volcanoes on Earth, bulldozing earlier deposits into deep channels. The extensive volcanic activity has built around a hundred and fifty mountains on Io, the tallest of which exceed Mount Everest in height.

Io's life is one of stress. Although imprisoned in the tight grip of Jupiter's gravitational field, its body is never at rest. It is always in fits, always fevered, always bleeding, always twisting, contorted like the people in the hellish paintings by Hieronymus Bosch.

Europa is the second-nearest Galilean satellite to Jupiter. By contrast to fiery Io, Europa is a world covered with cold ice, as smooth and completely spherical as a billiard ball. It is almost featureless. Only the nearly sunken traces of a few recent meteor craters break the monotony of a white, flat landscape. The ice is cracked into floes, with mineralised water splashing and oozing up through the gaps, overflowing onto the surface, staining it in a spider's web of red threads. The red stains are traces of deposits left by the evaporating water.

Europa looks like a world in stasis, but there is activity below its icy surface. The ice is a kilometre thick, floating on an ocean of salty water perhaps 5 kilometres (3 miles) deep. The water is warmed from below by geothermal energy. When the ice floes thrust together, they make hills of ice on the surface, but they are only a couple of hundred metres high. The landscape here is similar to the Arctic sea ice around the coasts of northern Canada or Siberia.

Altogether, there is more water on Europa than there is on Earth. In some future space mission, a lander might settle on the ice and try to penetrate through the ice layer, perhaps by using a radioactive probe to melt the ice and work its way down through the meltwater. It would be a suicide mission for the probe, as the water would refreeze above it, sealing it in. But what would it find? The still waters of Europa run deep, and it is tempting to imagine that, as it breaks through the lower surface of the ice, the penetrating probe might shine its lamps on alien oceanic creatures, taking pictures of them swimming below the maze of ice floes.

Ganymede is the largest of the Galilean moons, indeed the largest moon in the solar system. It is larger than Mercury, although only half as massive. Callisto is almost as big, the third-largest moon in the solar system, as big as Mercury but a third as massive. This means that they are both much less dense than planets made of rock and iron. The two moons must be mixed with something much lighter. That something is water – liquid water and ice.

Both Ganymede and Callisto have rocky surfaces that are cratered, like the Moon and Mercury. They look like the Moon, particularly Ganymede, which has two sorts of surface. A third of it is dark in colour, with lots of craters (thus, very old). The other two-thirds is lighter in colour, not so cratered (so younger): its peculiarity is that it is laced with grooves and ridges.

The lighter terrain on Ganymede is like the Moon's *maria*, caused by the upwelling of molten material from the interior that flooded lower areas of the surface. The difference is that the upwelling was not lava, but water melted by an asteroid impact. Callisto is similar, with parts of its surface lying in ripples, waves frozen by the intense cold.

Ganymede has an iron core that produces a weak magnetic field, but not Callisto. Their big secret is that several lines of argument point to both satellites having, under their rocky surface, a liquid ocean of salty water.

The ocean hidden in Ganymede is perhaps 1,000 kilometres (600 miles) deep, and like Europa's ocean holds as much as or more water than on Earth. Callisto's ocean is only a few hundred kilometres deep. Like Europa's oceans, these oceans (if they do indeed exist) may have life swimming in them.

There could well be a greater chance of finding life on the Galilean satellites than on Mars.

CHAPTER 11

Saturn: lord of the rings

- Scientific classification: *Gas giant.*
- Distance from the Sun: *9.54 times the Earth–Sun distance, 1,433.5 million km (890.8 million miles).*
- Orbital period: *29.5 years.*
- Diameter: *9.45 times Earth, 120,536 km (74,897 miles).*
- Rotation period: *10.2 hrs.*
- Average temperature of the top of the clouds: *–140 °C.*
- Secret excuse: *'That satellite and I had a very close relationship, but it broke up. At least I got a ring out of it.'*

The planet Saturn was named by the ancients after the Roman god of time, the same god as the Greek Titan Cronos. His name has given us words like 'chronometer'.

The association of the planet with the Olympian god of time is presumably connected with the fact that Saturn, as the furthest planet known in antiquity, was the slowest moving.

The planet has had a succession of secrets, uncovered one by one over a period of 400 years, concerning its most famous feature, its rings. But even now there remains one big secret; namely, what was the exciting event in its life that decorated Saturn in its glorious jewellery.

Saturn's rings are not completely unique in the solar system: there are rings around all the gas giants – Jupiter, Saturn, Uranus and Neptune. Indeed, it seems there are rings around three minor planets – Chariklo, Chiron and Haumea. These rings were discovered in the last fifty years using subtle clues. They are thin threads, hard to see, perhaps small bodies that have been broken up and now orbit their parent body. Saturn's rings are, in contrast, by far the most prominent of the ring systems in the solar system, and certainly the most complex and beautiful. Their dramatic appearance could be read as a fanfare proclaiming that they are important.

Saturn's rings have been known since Galileo turned his telescope to the planet in the seventeenth century, although their form remained unknown to him during his lifetime. He was confused by what he saw over the decade during which he observed Saturn's rings, because his telescopes were not clear enough to reveal their true shape.

At first, in 1610, he saw what he described as *ansae* ('handles'), as if Saturn was a drinking mug with two handles on opposite sides, like a loving cup. With the discovery of the moons of Jupiter in mind, he interpreted the extensions to the disc of Saturn as close, large moons: 'I have observed the highest planet [the most distant known to him, Saturn] to be tripled-bodied. To my very great amazement I saw that Saturn is not a single star, but three together, which almost touch each other.'

Two years later he expressed amazement that the moons had disappeared. 'I do not know what to say about so surprising a case, so unexpected and so novel.' Referring to the ghastly incident of infanticide and cannibalism by Cronos in classical mythology (which brings to mind Goya's terrifying painting of a deranged Cronos devouring his son), he asked rhetorically: 'Has Saturn swallowed his children?'

In 1616 he saw a more complex shape: 'The two companions are no longer two small perfectly round globes . . . but are much larger and no longer round . . . there are two half ellipses with two little dark triangles in the middle, each contiguous to the middle globe of Saturn, which is always perfectly round.'

The mystery of the changing appearance of Saturn was solved in 1656 by the Dutch astronomer Christiaan Huygens. The 'handles' were how a flat, inclined disc looked, centred on the planet. In the same way that Galileo published his discovery of the phases of Venus, Huygens published his discovery as an anagram, one into which he had not put much effort: aaaaaaa ccccc d eeeee h iiiiiii llll mm nnnnnnnnn oooo pp q rr s ttttt uuuuu. He later decoded it as: *Annulo cingitur, tenui, plano, nusquam cohaerente, ad eclipticam inclinato*, meaning, '[Saturn] is surrounded by a thin, flat ring, nowhere touching, and inclined to the ecliptic.'

The inclination of the rings to the orbits of Saturn and of the Earth is 27 degrees, and it is this fact that causes the rings to change appearance. When the Earth lies in the plane of the rings, they are edge-on and all but disappear because they are so thin, as Galileo reported in 1612. When the rings lie at their greatest angle, they elongate the overall image of Saturn into an ellipse, as seen by Galileo in 1616.

As telescopes improved, astronomers were able to pick out internal structure in the rings: gaps. The largest gap, first to be discovered, was found in 1675 by Giovanni Domenico Cassini. It separates what came to be termed the A- and the B-rings. Close-up pictures show further rings, separated by gaps wide and narrow. The individual rings have been labelled alphabetically in order of discovery. Saturn is 58,000 kilometres (36,000 miles) in radius and the closest ring to the planet is the D-ring, not much above Saturn at about 70,000 kilometres (43,500 miles) from Saturn's centre. The B- and A-rings are the brightest and the widest, lying between 90,000 (56,000 miles) and 135,000 kilometres (84,000 miles) from the planet, separated by what came to be called the Cassini Division, after its discoverer. Beyond the A-ring and another gap called the Encke Division is the F-ring, 140,000 kilometres (87,000 miles) from the planet. (The German astronomer Johann Encke did not discover this gap; it is named in his honour.) These inner rings all have a similar appearance, which makes it seem that they have a common origin.

The first interpretation of the rings was that they were a thin, monolithic, solid structure, like a vinyl record. As more and more gaps were found, the rings were envisaged as a collection of concentric, solid, hoop-like ringlets. In 1848, the French scientist Édouard Roche showed, however, that no large, solid structures, of whatever shape, that orbited so close to Saturn would survive. They would break up under the tidal forces from the planet. What this means is that the gravitational attraction on the nearest part of the structure to Saturn would be pulled hard and away from the most distant part, which would be pulled less. If the break-up force is greater than the internal strength of the structure, it would disintegrate.

The break-up force is greater the closer the structure orbits to the planet. The 'Roche limit' is the distance within which a solid satellite cannot survive. Comet Shoemaker-Levy 9 did not respect this limit when it ventured too close to Jupiter in 1992, and so it broke into more than twenty pieces. The Roche limit amounts to 2.44 times the radius of the planet, and all the main rings of Saturn lie within this limit, out to and including the F-ring. In 1857 the Scottish physicist James Clerk Maxwell showed that Saturn's rings could only be a large number of small, solid particles, orbiting independently.

The rings are very thin – perhaps 1 kilometre (0.5 mile) at their thickest, in places as thin as 10 metres (30 feet). To compare the rings to a vinyl record is to exaggerate their thickness relative to their diameter. To make the thickness the right scale in comparison with the diameter, the record would have to be thinner than a sheet of paper. The 'gaps' are not really empty spaces; they are zones where there are fewer particles than elsewhere. In fact, the rings themselves are made up of innumerable gaps and rings, like the grooves and bands into which a vinyl record is separated according to the musical content. The rings consist of particles 1 centimetre to 5 metres (0.5 inch to 15 feet) in size (pebbles to boulders), made primarily of water ice, and of micron-sized dust. There are millions of these particles and they jostle for room. The numerous collisions grind away the larger particles, increasing the amount of dust.

Saturn has over sixty individual moons, ranging in size from under a kilometre to the largest, which is bigger than Mercury. Given that the rings of Saturn are made up of innumerable small pieces, there is presumably a continuum between moons and ring particles. It is hard to fix a division between them: there must be plenty of moons that are smaller than 1 kilometre (0.5 mile) in size and have not been individually identified.

The majority of Saturn's larger moons orbit at high inclinations, in both directions: these are probably accidental satellites, asteroids captured by Saturn, having approached the planet from haphazard directions on disparate occasions. But twenty-four of the moons orbit Saturn in the same plane and in the same direction as the rings, some of them *within* the ring system. Atlas, Daphnis and Pan are three such small, inner moons. Atlas orbits near the outer edge of the A-ring, Daphnis inside the Keeler Gap in the A-ring, and Pan within the Encke Gap, also in the A-ring.

Saturn's moons play a critical part in controlling the rings, by pulling the individual particles of the rings away from some orbits and into others. This is what sorts the particles into rings and gaps. The process is called 'shepherding', as if the particles were sheep being chivvied by a sheepdog to take a particular route.

Sometimes the gaps are cleared out by a moon that plunges through the rings in a slightly eccentric orbit, pulling the particles off to one side or the other. Pan is one moon that shepherds the rings in this way: the moon is named after the god of shepherds. It is only 30 kilometres (18 miles) in size but it clears a gap 325 kilometres (200 miles) wide.

Pan was discovered after a theoretical prediction by US astronomers Jeffrey Cuzzi and Jeffrey Scargle in 1985, based on a discovery made when Cuzzi was whiling away time on a stopover in the airport in Albuquerque, New Mexico. He was shuffling through photographic paper prints of pictures taken by the *Voyager* space probes. Squinting along the surface of a print of Encke's Gap, he spotted in the foreshortened image that the gap had wavy edges. He realised that the wavy edges could be the result of a small moon orbiting within the gap. As each particle at the gap edge passes the moon, it receives a gravitational tug that pulls it into a more eccentric orbit. It then runs into the particles at the extremes of its orbit, which are moving in circular orbits. This creates the wavy pattern.

Back at NASA's Ames Research Center with his colleague Jeff Scargle, Cuzzi calculated how the waves were made, how big the satellite must be and where it would be in its orbit. It was too tedious to search through mountains of paper prints to confirm the theory by finding an image of the moon – seeing is believing – but five years later, in 1990, the archive of 30,000 pictures of Saturn made by *Voyager* was released in digital form on a CD-ROM.

Cuzzi had worked on the orbit of the undiscovered moon with a colleague, Mark Showalter. Showalter wrote a computer program to run through the archive, and identify and list all the pictures that had looked in the right place at the right time to see the moon. He left home for work one morning, saying that he was going to inspect all those pictures and find the moon. And – eureka – he did!

Just as Pan creates and sculpts the Encke Gap, so Daphnis sculpts the Keeler Gap, Atlas the A-ring. The phenomenon of shepherding is common in the Saturn ring system.

Atlas, Pan and Daphnis have been imaged in close-ups by *Cassini* in 2017, and have a curious shape. The clearest pictures are of Pan and Atlas. They are shaped somewhat like ravioli, with a central, white, smooth, spherical body circumscribed by a raised equatorial ridge, corresponding to the pinched edge of the pasta that forms a raviolo package. Perhaps it is unkind to say that they look like the frill of a tutu on an overweight ballerina.

It seems likely that the three satellites formed within Saturn's rings at a time when they were thicker, with ring material falling from every direction to give the central round shape. The rings became thinner and formed gaps. Residual ring material rained down on the satellites' equators. This built up as the equatorial ridges. Pan and Atlas are not very massive, so the ring material did not impact at speed: it just settled like snow, building up into a wall. Nor has the moons' force of gravity been strong enough to flatten out the ridge over time. Neither of them is at all dense – their average density is less than half that of water. This is the same as newly settled snow, which is an accumulation of ice crystals with spaces in between. It seems that the material of which Pan and Atlas are made is similar.

Saturn's moon Mimas is also very active in maintaining the structure of the rings and gaps, through a somewhat different mechanism. It is 200 kilometres (120 miles) in diameter and orbits not far outside the rings. Particles on the inner edge of the Cassini Division orbit exactly twice as fast as Mimas, and the repeated force on the particles drives them off to the side. This limits any tendency of particles in the A- and B-rings to fill the Cassini Division. Particles at the boundary between the

C- and B-rings are in a similar situation, orbiting three times as fast as Mimas. Prometheus is another shepherd satellite, this time for the inner edge of Saturn's F-ring.

The pebbles in Saturn's rings populate the space around Saturn with what amount to test particles that show how the space is filled with gravitational forces due to the planet and its system of satellites. This has turned out to be an amazingly rich situation for astronomers to study. The theory of gravity is 300 years old and, before these analyses of Saturn's rings, everyone thought its subtleties were all understood. However, there were more secrets to be uncovered, exposed in the behaviour of the rings.

In an additional complication, *Voyager 1* discovered dark, nearly radial 'spokes' in the B-ring, which are about 8,000 kilometres (5,000 miles) long and 2,000 kilometres (1,200 miles) wide. They develop over minutes, rotate with the rings and disappear in a few hours; they also come and go over years, perhaps in a way correlated with Saturn's orbit. The spokes are unexplained but seem to be dust gathered and held in place by electrostatic forces.

The biggest secret of all, and one that is still not known for sure, is where the rings come from. For a long time, based on Roche's work, people thought that the particles were the result of the break-up of a Mimas-sized satellite that ventured too close to Saturn and was broken apart by the planet's tidal forces. Other people think the rings are as old as Saturn itself, formed as part of its birth process. A third hypothesis is that a comet encountered Saturn and broke up. This would account for the composition of the ring material.

Cassini provided a clue that the birth event was relatively recent. The space probe spiralled under the rings into Saturn's atmosphere in its last few days, its so-called Grand Finale manoeuvre. It was a work programme that was too risky to execute during the main mission, for fear of a collision between the spacecraft and a rock that had strayed outside the rings. Cruising inside the B-ring, nearest to Saturn, *Cassini* encountered an unexpectedly strong 'rain' of ice and other simple chemicals falling from the rings onto the planet. The rings are disappearing quickly, which suggests they cannot be too old.

Saturn mimics Jupiter, but it is not as large, is further from the Sun and colder. As a result, Saturn has a less exciting existence in itself, with less weather and fewer violent storms. Its structure is similar to Jupiter's, with a rocky core, surrounded by metallic and then liquid hydrogen, and then by hydrogen and helium gas. There appears to be less helium on Saturn than on Jupiter, but that is thought to be because the helium has sunk below the cloud tops and cannot be seen, not that there really is less. As with Jupiter, the gas at the cloud tops is mixed with chemical impurities, which make Saturn a pale yellow: crystals of ammonia are thought to be responsible for this colour. Other chemicals include acetylene, ethane, propane, phosphine and methane, ammonium hydrosulphide and water. There are shades of yellow in bands, and hard-to-see circular swirls and storms.

Saturn rotates only a little slower than Jupiter, and has a similar ellipsoidal shape. The core of Saturn rotates with a period of ten hours and thirty-three minutes, as measured by the cyclic nature of its radio emissions, which are fixed in its magnetosphere, which is fixed to its core. The winds of Saturn are some of the fastest in the solar system and affect the apparent rotation speed of the cloud tops. There is an enormous difference between the rotation period near the poles (10 hrs 40 mins) and the equator (10 hrs 15 mins).

Although the winds on Saturn stream so quickly around the planet, the turbulence in Saturn's climate and weather systems is less pronounced than Jupiter's, presumably because Saturn is twice as far from the Sun as Jupiter and correspondingly colder, and even though Saturn's axial tilt is 27 degrees, similar to the Earth's. The contrast of the weather patterns on Saturn's cloud tops is thus less pronounced than on Jupiter, and the planet is less interesting to view in this respect. There are some exceptions to its bland appearance: Saturn has a Great White Spot, a name that bigs up this phenomenon to make it seem as important as Jupiter's Great Red Spot. It is a storm that appears every thirty years or so, once per orbit. It is triggered when Saturn's north pole is tilted towards the Sun. In another case, in 2004, the *Cassini* spacecraft saw a convoluted, swirling cloud feature called the Dragon Storm. It generated bursts of radio waves, and has been interpreted as a giant thunderstorm, with the radio bursts produced by lightning.

There is a unique feature of Saturn's atmosphere, which was unsuspected before the Space Age. Because from Earth we do not see the poles of Saturn clearly, it was not until the *Voyager* missions in 1981 had a close-up view, confirmed by the *Cassini* mission in 2006, that space scientists saw an extraordinary hexagon of clouds at Saturn's north pole. It is a feature unique in the solar system. The sides of the hexagon are about 14,500 kilometres (9,000 miles) long; its area would easily encompass the Earth several times over. The north pole has been imaged through various coloured filters on several occasions. The hexagon appears in all of these images. The significance is that different colours come from different depths in the atmosphere of Saturn. This suggests that the body of the hexagon is hundreds of kilometres high. Lots of purported explanations for the hexagon shape have been put forward but, so far, the jury is out and the reason for the precise geometrical shape remains one of Saturn's secrets.

The argument about this newly discovered structure exemplifies one of the reasons for exploration of the solar system. Meteorological science has been developed to be able to predict Earth's atmosphere – its weather and climate. The atmospheres of other planets challenge meteorological science in new ways and drive its development so that it attains new depths of understanding, which in turn feed on into terrestrial meteorology, giving it greater scope and accuracy. Just as the biographies of people provide insight into human character, from which we may draw lessons about our own behaviour, so the biographies of the planets provide insight into our own world, and the way that it impinges on us. Space scientists may look at and study the lives of other planets in the solar system, but they have our own planet in the back of their minds.

CHAPTER 12

Titan: animation suspended

- Scientific classification: *Satellite of Saturn.*
- Distance from Saturn: *1,221,850 km (759,220 miles), 3.27 times the Earth–Moon distance.*
- Orbital period: *15.9 days.*
- Diameter: *5,150 km (3,200 miles), 1.48 times the Moon.*
- Rotation period: *Synchronous.*
- Average surface temperature: *–180 °C.*
- Secret plan: *'If I can find the energy, I will put life into this neighbourhood one day.'*

Astronomers who study galaxies have the ability to peer back into the past by looking at distant ones. Light travels at a finite speed and, after it is emitted from a galaxy, takes some time to get to us. It carries news to astronomers about the way distant galaxies used to be when the light left. That time can be long in the past if the galaxy is far away. It is thus almost routine for astronomers to study galaxies in every stage of life. In biological terms, this is like studying children in a school, shoppers in a mall and elderly people in a retirement home, and inferring from this the way in which people age. Communities of people are separated by distance in a city, but galaxies are separated in the Universe by light years that are equivalent to time. The span of time available to astronomers to look back on in this way stretches back over 90 per cent of the age of the Universe, let us say 12 billion years or so.

Astronomers who study planets, on the other hand, are not in such a good position. Planets in the solar system are far away on the scale of our own environment, but not as far away as distances in astronomy go. It takes just one hour and twenty minutes for light to travel from Saturn to the Earth. It is no help in working out the history of a planet to know what it was like an hour ago, given that its age is 4.6 billion years: an hour is almost no time at all.

There is one sense in which astronomers can look back into the past when studying Saturn. Saturn is on the outer reaches of the solar system, where the gravitational influence of the Sun is weak and its light is dim. As a result, Saturn is cold. Moreover, at this distance from the Sun, asteroids are fewer in number and move less quickly. There is altogether less going on out there. In the cold, chemical processes are less vigorous, and in the weaker gravity there are fewer and slower collisions. If there is an Earth-like world in the outer reaches of the solar system, it is likely to be less advanced in its evolution than the Earth is. It could be like the Earth soon after it was formed, perhaps even before life evolved here.

Although astronomers cannot in reality travel in time back to the distant past to see what the Earth actually looked like, the solar system has given us a lucky example of a world that is like the Earth as it used to be. The example is Saturn's satellite Titan. By sending robotic spacecraft there to study it, astronomers have travelled in distance what they couldn't travel in time: back towards life's origins.

Saturn and its satellites were first explored by spacecraft in brief fly-bys: by *Pioneer 11* in 1979 and the two *Voyager* spacecraft in 1980–1. In 1997 the NASA/ESA *Cassini-Huygens* spacecraft was launched from Cape Kennedy in Florida to study the Saturn system. On arrival at Saturn, the mission separated into two spacecraft. A space probe called *Huygens* parachuted onto Saturn's largest satellite, Titan, while the *Cassini* orbiter travelled in and through the Saturnian system from 2004 until 2017. This joint mission – one of the most successful planetary exploration missions ever – has transformed our knowledge of Saturn. It opened our eyes to the potential for the development of life in places that were previously thought completely impossible.

I watched the launch of *Cassini-Huygens* at night from the Cape Kennedy spaceport. I stood beside a drainage canal near the launch pad. Not far away, down a vertical bank, an alligator looked up at me, blinking. Its eyes sparkled in the moonlight and its scales glistened from the water that sloshed over its back when it writhed in the mud. A Titan IVB/Centaur rocket, the most powerful rocket available at that time, stood on the launch pad, under floodlighting. It was a thousand tonnes of brute force with two tonnes of sophisticated twentieth-century space technology on board. Somewhere near the launch pad, in a bunker safe from accidents, the mission controllers were ticking off the countdown that would send the spacecraft to places separated from me by an enormous distance. As it proved, the places were also separated from me by eons of time. If there was life in the Saturnian system, it was much more primitive compared to the alligator than the alligator was compared to me. The rocket took off with a roar and a blaze of light from its burning fuel, arcing away over the Atlantic Ocean, climbing up into space. I waited the seven years it took to get to Saturn before hearing about its first discoveries.

Saturn's Titan is only about 100 kilometres (60 miles) smaller than Jupiter's Ganymede, but bigger than the planet Mercury. It is so large that it retains a thick atmosphere, the only satellite to do so: atmospheres on other satellites are tenuous and temporary, if they have atmospheres at all. From Earth, our telescopes see Titan as a featureless sphere, just uniform cloud tops. Sensors on the first fly-bys of the space programme could not penetrate through the cloud and saw no visible surface features. What they did see and investigate, while passing near to the moon, was its dense atmosphere, an orange haze when backlit by the Sun. The atmosphere of Titan extends nearly 1,000 kilometres (500 miles) into space.

The atmosphere is mostly nitrogen, but a few per cent (1–5 per cent depending on atmospheric height) is methane, with traces of helium and argon, and other hydrocarbons. The haze is a smog of sooty particles, produced by the action of the Sun's ultraviolet light on the methane.

The continued existence of methane in the atmosphere of Titan is significant, since the action of sunlight should convert the entire atmosphere to other hydrocarbons within 50 million years. There must be a source of methane on Titan itself: a large reservoir, volcanic vents, or even, hypothetically, biological activity.

203

The surface temperature of Titan is −170 °C and the atmospheric pressure at the surface is 1.5 times that at the surface of the Earth. Under such conditions, methane condenses to a liquid. Titan must be 'wet' – not with water but with liquid methane. Until the *Cassini-Huygens* mission the nature of the wet surface was unknown. Planetologists wondered whether it is wet everywhere: a world-wide methane ocean? Is it wet with ponds and lakes among rocks? Is it wet and soggy like a marsh? The arguments were gone into in detail during the design of the *Huygens* lander: the answer obviously would make a difference to the survival of the space probe. It was an argument that remained unresolved until *Huygens* went there. This secret of Titan's was revealed only at the moment of touchdown.

Huygens was carried piggy-back by *Cassini* to Saturn and dropped off onto Titan. The journey to Saturn was a long seven-year spaceflight, with *Huygens* dormant for most of the time, woken up briefly every six months for a health check. It was quite an achievement to make equipment that could be stored for seven years in space conditions and have it work when it was needed. The lander was going to descend to its target on parachutes whose material had to be tightly folded so as to fit inside the spacecraft – would the parachutes unfurl? *Cassini* was powered by a radioactive electricity generator but *Huygens* had batteries. The batteries had to leak as little of their charge as possible so that they had enough left to operate electrical machinery (sunlight at Saturn is so weak that solar panels would be ineffectual). The computers had to be programmed in software that had been more than usually tested but therefore was already old when it left Earth – would there be mission controllers who understood the computing language at the time of descent so that they could effect any necessary changes? In fact, would any of the people who made the equipment be around to tell controllers how it worked and what to do if there was a problem? The answers were yes, thanks to the project discipline with which the mission was planned and controlled at NASA and the European Space Agency.

Cassini arrived successfully at Saturn on Christmas Day, 2004. It separated the *Huygens* lander by exploding the bolts that held it on and releasing springs that thrust the lander into space. For two weeks, the lander navigated its way to its target, falling and then parachuting onto Titan. The lander's scientific payload was powered with batteries that lasted for somewhat over three hours. The slow descent took two of them, during which time the spacecraft determined the composition of Titan's atmosphere and made other measurements. The lander swung like a pendulum below the parachutes, slowly spinning, drifting in the wind. It was entirely autonomous during its descent. If there had been any decision-making to do, it was no use radioing back to Earth and asking controllers for help. By the time radio waves had travelled back to Earth and got an answer, at least three hours would have passed and the problem would have solved itself, or otherwise.

A camera on the lander documented the view during the descent. It headed towards a rocky shoreline. There was a flat plain that abutted a region of hills cut by drainage channels – rivers. On which side of the shoreline would Huygens land? Would it land in the hills and tumble over, down the side of a valley? Would it land safely in the flat area? What was the nature of the flat area? Was the lander about to drop onto a stable rock surface? Would the lander sink out of sight in a lake or be swallowed by quicksand? The first thing that *Huygens* was going to do to explore the surface of Titan was to touch it. A thin probe extended below the lander, designed to see how hard the landing place was – the arguments during the design process suggested that it might land with a splash, a bang or a squelch.

It landed with a soft thump, although no one was there to hear. The landing site was relatively smooth, but not a liquid. The surface was neither hard and solid nor soft and fluffy; it was slightly compressible, like lightly packed snow or wet sand. The weight of the lander settled it gradually by a few millimetres, pressing a pebble into the sand under its foot as it finally grounded.

The camera on the lander recorded the area around the landing site, and had time to transmit one picture up to *Cassini* and back to Earth. It showed rounded boulders, which had rolled down the rivers into an estuary that flowed out onto the wet sand. The picture is in some ways prosaic, a commonplace landscape that can be seen on every muddy shoreline on Earth that is broken up by outflowing rivers. It is a landscape one might visit on an invigorating walk on a public holiday weekend, but not one that is attractive enough for a lazy holiday. But behind the picture of this terrestrial-looking scene is something like nothing on Earth. The landscape was created not by water but by liquid methane. On Titan, methane rain falls on the hills and liquid methane streams down the rivers, carrying ice floes to be deposited onto the drying bed of a methane lake.

The surface of Titan is hidden from view from outside by its smoggy, opaque atmosphere, but as the *Cassini* mission progressed, orbiting around Saturn and making several passes close to Titan, it was able to use radar to penetrate through the smog and survey the entire surface. The *Huygens* landing site was one place in a cosmic liquidscape, a mosaic of small irregularly shaped methane lakes. Elsewhere there is a 400-kilometre-(250-mile)-long river flowing through an area of steep-sided canyons up to 600 metres (2,000 feet) deep. One picture of Titan taken by *Cassini* looking back towards the Sun across one of the lakes shows the evening glint of the setting Sun reflected under the atmosphere. When the winds are high, the lakes have high waves rolling slowly across their surface. (Titan might be a good place for surfing.) The lakes progressively dry out and then refill with the progression of the seasons on the satellite. *Huygens* landed where it did at a lucky time, in the dry season. If it had landed at the wrong time, it could have splashed into the lake and sunk.

The atmosphere of Titan is like the Earth's atmosphere as it used to be, and the rich carbon chemistry of its atmosphere and lakes is thought to resemble the carbon chemistry that preceded life on Earth. The chemical ingredients for life are there in Titan's pre-biotic atmosphere. There is no evidence for life itself, although perhaps some archaea (primitive bacteria-like organisms) live at the lake shores. They certainly have not provoked a Great Oxygenation Event on Titan like the one that gave Earth's atmosphere its oxygen gas about 2.4 billion years ago. We know that has not happened because oxygen would combine with the methane and remove it from the atmosphere. Future spacecraft in the form of drones might explore Titan by flying in its atmosphere to look for life in the methane lakes. Will Titan prove definitely to be pre-biotic or will it show itself to be positioned right at the moment life starts?

CHAPTER 13

Enceladus: warm hearted

● Scientific classification: *Satellite of Saturn.*

● Distance from Saturn: *238,000 km (149,000 miles),*
0.62 times the Earth–Moon distance.

● Orbital period: *1.37 days.*

● Diameter: *500 km (310 miles), 0.145 times the Moon*

● Rotation period: *Synchronous.*

● Average surface temperature: *–198 °C.*

● Secret pride: *'I may look cold on the surface but I*
have a warm heart.'

Saturn's moon Enceladus is a small, rocky and icy sphere
500 kilometres (310 miles) in diameter. Its nature arrests
attention much more than its size. In some ways, Saturn's
Enceladus is a cousin to Jupiter's Io. Io has volcanoes, and

so does Enceladus – not fiery volcanoes erupting with hot lava, but cryovolcanoes ('cold volcanoes') erupting with geysers of icy water. The water falls as snow across half the satellite. Enceladus is a cross between Yellowstone Park in Wyoming and the ski resort of Aspen in Colorado.

The existence of the cryovolcanoes was uncovered through a chance observation by a team led by Michele Dougherty of Imperial College, London. The instruments on a scientific satellite are made by a team led by a principal scientist, the one who takes responsibility for delivering it working and on time, and warrants not only its performance but that it will not cause trouble for the other parts of the satellite. Dougherty was the principal scientist for an instrument on the *Cassini* spacecraft that was used to map Saturn's magnetic field. Wherever the spacecraft went, the instrument made a measurement. In 2005, *Cassini* was flying by Enceladus at a considerable distance, and Dougherty and her team had no expectation of seeing anything significant. They were unexcited enough by the prospect of what they assumed would be a routine measurement that they did not even look at the data for a day or two. However, when they did, they noticed that a disturbance in Saturn's magnetic field was being dragged along by Enceladus. This suggested that the moon had some sort of atmosphere that had trapped the magnetic field and was pulling it along. Dougherty saw the same thing as the probe made another pass by Enceladus. There were some indications that the 'atmosphere' was made of water. Enceladus is too small to have a permanent atmosphere and it would be rare for a moon to have an atmosphere of water, except briefly, after a comet's impact: what was going on?

Dougherty and her team sat on the discovery for some weeks, working and reworking the data to make sure of what they had seen, looking at the implications. Presenting the discovery at a conference about the mission, she was able to persuade the mission controllers to fly through the affected area to confirm that there was material there. It involved a flight by the probe very close to the surface of the moon. The mission controllers were excited by the prospect of breaking the then record for the closest pass made by a space probe over a moon, just 173 kilometres (107 miles) above the surface – it would be a feat they could boast about. The scientists at first were sceptical about any change to the mission programme that they had honed over the past several years. But they became convinced that the discovery would be important and *Cassini* should try to get it nailed down. Later, the scientists and the controllers flew the probe at a height of only 25 kilometres (15 miles), where the 'atmosphere' was so dense that the spacecraft was on the verge of tumbling out of control – that was almost too much bravado!

As the data built up, it became clear that the 'atmosphere' was localised at the south pole of Enceladus. From the region known as the Tiger Stripes, the moon produces sprays of water vapour and water-ice chips (hail and snow), in a gas of methane, carbon dioxide and other simple organic molecules. Seeing is believing, again, and the sprays were pictured as fountains on a pass of the spacecraft over the moon's surface in 2006. It had been positioned specifically to view the sprays, backlit by the Sun.

The total amount of ice that is vented into space around Enceladus is roughly the same as that from the Old Faithful geyser at Yellowstone. Some of the ice crystals fall down towards Enceladus and some are ejected into space and feed its E-ring. This diffuse ring lies outside the main rings of Saturn. The ring outlines the orbit of Enceladus, rendering the orbit visible, curving in space around the planet.

Just as the surface of Io is covered by sulphur and ash from its volcanoes, half the surface of Enceladus is covered by ice from its cryovolcanoes. The terrain that covers the north pole of Enceladus is old and heavily cratered like our own Moon. The craters are distorted, eroded and cut by chasms – there has evidently been considerable geological activity since the craters were formed.

By contrast, the southern hemisphere of Enceladus, where the geysers are, is new: it is smooth, slightly wrinkled terrain, coated by the sprays of snow and hail. The snow falls back from the sprays onto the surface of Enceladus. Snowfall over millions of years has blanketed areas of the surface in a thick layer. The tiny snowflakes have blanketed the rocky surface of Enceladus, smoothing out the hills and depressions. Some of the more pronounced features of the landscape still show on the snow's surface, like ghosts: old buried craters and canyons, the largest of them comparable in size to the Grand Canyon of Arizona.

The overlying layer of fine, powdered snow is sometimes 100 metres (300 feet) deep in this area. It is considerably deeper than the snow at a ski resort. But it accumulated at a rate that is very much slower than the snowfall the management of a ski resort would wish to happen at the start of the winter season. The rate at which the surface builds up is less than a thousandth of a millimetre per year. Over millions of years, even at this slow rate, it has made a fine *piste* – the skiing surface is permanent and guaranteed! The rollers on the lakes of Titan make it a good place for surfing, and Enceladus is the place to go to ski – it would be expensive to travel to the Saturnian system for a holiday but there are few places that combine opportunities for summer and winter sports so well.

The geysers on Enceladus that have produced this potential all-year playground for winter sports are fed by reservoirs of liquid water not far below the surface. The land is cut by large, parallel, dark cracks, the Tiger Stripes, in the depths of which are warm areas. The warm rocks are heated by the flexing of the body of the moon by the tidal forces of Saturn, in much the same way that Io is flexed by Jupiter. Inside Enceladus, therefore, is hot rock, whose warmth melts its internal ice and fills its underground caverns with water, laced with organic chemicals in solution. The underground water reservoirs are large: the top of the sub-surface ocean is at a depth of perhaps 30 kilometres (20 miles) under the surface and may be as much as 10 kilometres (6 miles) deep.

The environment in this ocean is similar to some niche environments on Earth: wet, warm, dark caverns, deep within volcanic rocks. The implication is that Enceladus is a potential habitat for life. When Charles Darwin was writing about where life might have originated on Earth, he envisaged that it might have started in a 'warm, little pond' (Chapter 2). On Enceladus, life might have started in a 'warm, giant cistern'. This makes Enceladus a potential target to explore in the search for extraterrestrial life. It might be the easiest place to go to do this. The geysers on Enceladus bring samples of once-warm water above the surface of the moon, where they could be collected for analysis by a spacecraft searching for extraterrestrial life. It could fly through the spray and would not have to bore through a kilometre of ice as it would if it looked for life in the sub-surface ocean on Jupiter's moon Europa. It doesn't even have to land on Enceladus to see if there is life there. No wonder that astrobiologists are drawn to study this moon, dreaming about sending future missions to explore it, to uncover its remaining secrets!

CHAPTER 14

Uranus: bowled over

> ◉ Scientific classification: *Ice giant.*
> ◉ Distance from the Sun: *19.2 times the Earth–Sun distance, 2,872.5 million km (1,784.8 million miles).*
> ◉ Orbital period: *84.1 years.*
> ◉ Diameter: *4.01 times Earth, 51,118 km (31,763 miles).*
> ◉ Rotation period: *17.9 hrs.*
> ◉ Average temperature of the top of the clouds: *−165 °C.*
> ◉ Secret power: *'I have a completely different perspective on the Universe.'*

Uranus was unknown in antiquity. In principle, it can be seen under the most favourable circumstances by the naked eye, but not at all readily, so it is no surprise that no one noticed it before telescopes were invented. It was, in 1781, the first planet to be discovered. Its very existence appears to extend a curious formula called Bode's law about the distances of the planets from the Sun, which seems to describe something significant about the architecture of the solar system. If only we could understand this secret that had been uncovered, everybody thought, it would be an important scientific revelation. Some continue to hope that there is a scientific secret hidden there to be uncovered, but, if so, scientists have yet to find out what that is, if anything.

Other astronomers are sceptical. Their doubts are expressed, by coincidence, in the movement of Gustav Holst's musical suite, *The Planets*, that is devoted to Uranus. He subtitled the movement 'the Magician'. It has several magic tricks in its music, including an incantation at its final climax in which Uranus seems to become enveloped in flames and disappear, doubtless an illusion. Reluctantly, many astronomers have concluded that, like that magician's trick, there is in Bode's law less significance than meets the eye. It is no illusion, however, that Uranus is upside down.

The astronomer who found Uranus was William Herschel, working with the aid of his sister Caroline. In 1781 William was not actually an astronomer, but a musician with a curiosity about astronomy. He was born in 1738 in Hanover, and became a military bandsman. He fought as part of the British army at the Battle of Hastenbeck, then left the army; some untruthfully say he deserted it in the disorder that followed after it was defeated by the French. Whether this allegation is true or not, he certainly fled to England. He settled in Bath where he established himself as a music teacher and a church organist. It seems he was seen as an eligible bachelor by the society ladies who came for piano lessons. His sister, Caroline, also escaped from Hanover, not from an army but from her and William's abusive older brother, Jacob, who, exercising coercive control, kept her in thrall as a housekeeper. She had been scarred by smallpox, and was told by her family that, with a face like that, she would never attract a husband so should make her family the focus of her attention. She had resigned herself to that self-fulfilling, self-serving prediction, but did not see herself sewing stockings for her brother for the rest of her life. She succeeded in joining her much friendlier brother, William, in Bath, defending him against predatory widows, accompanying him by singing in his concerts, and working with him on their studies. She taught herself astronomy when William became interested in it.

William made telescopes, casting and grinding the mirrors himself in the basement of his house (it is now a museum: you can still see the paved floor there, cracked from the heat of a fiery accident while casting mirrors). He designed and formed the telescope tubes from wood and tin, and erected the telescopes on the garden lawn, or even in the street outside his house if that was what was needed to get a better view. His telescopes were the best of his time, with sharp optics, held steady on a sturdy mount, convenient to use, and they later on formed the basis of a profitable business.

William formed the idea to 'review' the entire sky, surveying every star and the spaces in between by letting the sky drift through the field of view of his telescope in parallel strips while he watched unwaveringly. Caroline kept everything systematic. They recorded double stars, star clusters and nebulae, compiling catalogues that became the broad framework that enabled later, detailed investigations for over a century afterwards.

On 13 March 1781, William saw a star worth a special note. It was the quality of the optics of his telescope that enabled him to see that there was something odd about its appearance. It was a 'curious either nebulous star or perhaps a comet'. Returning to view it over the next hours and days, the two of them found that it had moved and could not be a star, which would remain fixed. A comet, then? No, there were inconsistencies with that idea. A comet would likely be in a highly eccentric orbit crossing the solar system, but the curious object proved to be in a near-circular orbit, like a planet orbiting beyond Saturn. Also, comets have a fuzzy look, like hair, a so-called *coma*, and they often have a tail. The 'curious' object as seen in his telescope had the shape of a circular disc, as a planet would. If a bird looks like a duck and quacks like a duck, then perhaps it is a duck. The curious object looked like a planet and it behaved like a planet: it proved to be a planet.

William was invited to tell King George III of the discovery and then asked to make a telescope for Windsor Castle in order to show astronomical sights to the court, such as newly discovered comets. He was appointed as the King's Royal Astronomer and given a stipend so that he could be free to work full time – his sister Caroline, his co-worker, was given a stipend too. Her stipend was only half as much as William's, a not unfamiliar gender discrepancy even today. Be that as it may, Caroline was pleased with her money because it gave her a freedom that she had never before experienced: 'In October 1787 I received the first quarterly instalment of 12 pounds 10 shillings. It was the first money in all my lifetime that, at the age of 37 years, I ever thought myself at liberty to spend to my own liking.'

There was a certain amount of fuss about the name of the planet, William ingratiating himself with the British king but annoying non-British astronomers by wanting to christen it *Georgium Sidus* – the 'Georgian planet'. Eventually, it was the suggestion by the German astronomer Johann Bode that prevailed. 'We had better stick to mythology,' opined Bode, recommending that the planet should be named Uranus, after the Greek god of the sky. It is the only planet whose name derives directly from Greek mythology, the others being named after the gods of Rome.

Bode was a leader of German astronomy in the late seventeenth century and played a key role in uncovering and promulgating what has become known as the Titius–Bode law of the distances of the planets from the Sun, in the development of which Uranus played a significant part. Johann Daniel Titius was Professor of Physics at Wittenberg from 1756, and translated from French into German a work called *Contemplation de la Nature* by the Swiss scientist Charles Bonnet. Titius added into the text ideas of his own. Bonnet said in one passage that '[w]e know seventeen planets [and satellites] that enter into the composition of our solar system; but we are not sure that there are no more', and went on to anticipate more discoveries as telescopes improved. Titius then inserted what we now call the Titius–Bode law:

For once pay attention to the width of the planets from each other and notice that they are distant from each other almost in proportion to their bodily heights increase. Given the distance from the Sun to Saturn as 100 units; then Mercury is distant 4 such units from the Sun, Venus 4 + 3 = 7 of the same, the Earth 4 + 6 = 10, Mars 4 + 12 = 16 ... From Mars follows a place 4 + 24 = 28 such units, where at present neither a chief nor a neighbouring planet is to be seen ... Above this, to us unrevealed, position arises Jupiter's domain of 4 + 48 = 52; and Saturn's at 4 + 96 = 100 units. What a praiseworthy relation!

Bode read Titius's translation of Bonnet's book and he put the relationship as proposed by Titius into the text of his own book, an introduction to astronomy published in 1772, *Anleitung zur Kenntniss des gestirnten Himmels* (*Introduction to Knowledge of the Starry Heavens*). Although Bode is obviously following Titius, he does not even mention his name. It was his book that made the relationship interesting to other scientists; as a result, it became known as Bode's law. Titius's part in the story was rediscovered later and his name was rightfully attached, as the Titius–Bode law.

Here is the Titius–Bode law in tabular form, showing how, except for a gap, it proceeds almost solely by doubling up the distance of each planet from the Sun:

THE TITIUS–BODE LAW

Planet				Calculated distance	Actual distance
Mercury	0	+ 4	=	4	3.9
Venus	3	+ 4	=	7	7.2
Earth	6	+ 4	=	10	10
Mars	12	+ 4	=	16	15
The gap	24	+ 4	=	28	–
Jupiter	48	+ 4	=	52	52
Saturn	96	+ 4	=	100	95

Uranus added a further line to the table, which was an amazingly good fit:

Uranus	192	+ 4	=	196	192

The accurate fit of Uranus to the Titius–Bode law seemed to make the law more than a coincidence. There was more to come. A few years after the discovery of Uranus, the asteroid or dwarf planet Ceres was discovered. It fitted the gap between Mars and Jupiter (see Chapter 8).

The law seemed to have predictive power. Similar laws were discovered for the spacing of the four main moons of Jupiter in orbit around their planet, and for the large moons of Uranus.

The Titius-Bode does not do so well for Neptune.

Neptune	384	+ 4	=	388	301

However, there is a variation on the Titius–Bode law that works for the five planets orbiting the extrasolar planetary system 55 Cancri, and there is a more complicated generalisation that seems to fit a total of sixty-eight extrasolar planetary systems that have four planets or more. Of course, the more complicated a mathematical formula is, the more easily it can be adjusted to fit data accurately, without having any underlying basis for doing so.

Astronomers have looked for the origin of the law in some real phenomenon in the formation and life of the solar system. But nobody has ever found what that is. Maybe the interactions between the planets as exemplified in the Nice Simulation have something to do with it – the first presentation of the Nice Simulation that I ever saw provoked one astronomer in the audience to an ecstatic proclamation that the simulation's creator, Alessandro Morbidelli, had solved the secret of the Titius–Bode law at last! Morbidelli, however, disowned the possibility.

The Titius–Bode law might be the first appearance of something significant but hidden; or it might be a meaningless numerological curiosity. The prospect that it may be the result of something important is reminiscent of earlier episodes in the history of planetary motions. The German astronomer Johannes Kepler found a coincidence that related the spacing of the planets to the sizes of the five regular polyhedral solids nested inside one another. These solids are known as the Platonic solids and are the tetrahedron, cube, octahedron, dodecahedron and the icosahedron.

Kepler recorded the moment when he had his brainwave. On 19 July 1595, he was preparing to teach a class in geometry. He drew a circle on a blackboard, within which he drew a large number of equilateral triangles, with their corners on the circle. Within all these triangles appeared another, smaller circle, which touched the triangles' sides. Kepler suddenly realised that the ratio of size of the two circles was the same as the ratio of the size of the orbits of Jupiter and Saturn.

He went on to wonder whether he could fit the orbits of other planets in a similar way. He tried other planar geometric figures – a triangle, a square, a pentagon, and so on. This did not work out. Maybe three-dimensional geometric solids would be better, more representative of the planets as three-dimensional worlds.

In a model of the solar system, Kepler constructed a series of nested solids, somewhat like a Russian doll, working inwards from the orbit of Saturn, represented by a sphere. He fitted a cube inside, with its corners touching that sphere, and inside the cube he fitted another sphere that touched the sides of the cube. This sphere represented the orbit of Jupiter. Inside that sphere he fitted a tetrahedron. The sphere inside the tetrahedron represented the orbit of Mars. Inside that sphere he fitted a dodecahedron (Earth), followed by an icosahedron (Venus), and, finally, an octahedron with a sphere inside that represented the orbit of Mercury.

Kepler had inherited astrological and alchemical leanings from his mother, a woman who was once tried as a witch. The model of the solar system that he had stumbled across was quite a good fit and it had arcane properties that were appealing to him. He wrote a book on it, published in 1596, called *Forerunner of the Cosmological Essays, which contains the Secret of the Universe; on the marvellous proportion of the Celestial Spheres, and on the true and particular causes of the number, magnitude, and periodic motions of the Heavens; established by means of the five Regular Geometric Solids*. The title makes it clear that Kepler thought the coincidence was the key to the secret lives of the planets in orbit in the solar system. Human beings are inclined to find meaning where none exists, in the same way that some people see significance in the lucky numbers that turn up in a lottery ticket. The coincidence is just that: a coincidence with no fundamental significance.

Kepler, however, was a convinced mystic and continued to seek similar numerical relationships in the way the planets move. In 1619 he published what came to be known as

Kepler's Third Law relating the cube of the size of planetary orbits to the square of their period of revolution around the Sun. This proved to be a consequence of Newton's Law of Gravitation, and came about because the force of gravity between two bodies (like a planet and the Sun) is proportional to the inverse square of the distance between them. The meaningless numerological coincidence about the five geometric solids led Kepler to a significant discovery, behind which was a fundamental scientific law of nature. The forlorn hope was, and to a certain extent remains, that the Titius–Bode law might do the same. It is such an attractive pull for armchair theorists that the planetary science journal *Icarus* has had to put a brake on publication of the many submissions that it receives about it.

Uranus's orbit has properties that hint at secrets; Uranus itself is similar. Both Uranus and Neptune are less studied than the other planets, being so distant and having been less explored by spacecraft. In fact, both planets have been visited only once, in the late 1980s. The *Voyager 2* interplanetary probe visited Uranus in 1986. No other spacecraft visits are in the offing.

Superficially, Uranus mimics Jupiter and Saturn, but it is smaller than both, further from the Sun and even colder. As a result, Uranus has an even less exciting life than Saturn, and its cloud tops are almost completely uniform and featureless. But not quite. Uranus has a characteristic blue-green appearance due to a high layer of clouds of methane ice. Unlike Jupiter and Saturn its interior is composed of various ices rather than hydrogen and helium – sometimes it is termed an ice giant, rather than a gas giant. It also experiences large storms from time to time; nobody knows how they are triggered, but the suspicion is that they are seasonal.

Uranus has a weird magnetic field, strong but messy. It is not centred in the middle of the planet and it is tilted at an angle that does not line up with the planet's rotation. It is fifty times the strength of the Earth's magnetic field.

Uranus has quite a large retinue of more than two dozen moons. They are all named after characters in plays by William Shakespeare and a poem by Alexander Pope. The largest of them are considerably smaller worlds than the moons of Jupiter or Saturn, or indeed our own Moon. Miranda, Ariel, Umbriel, Titania and Oberon are the five largest moons of Uranus, up to 1,500 kilometres (900 miles) in diameter. The planet also has a ring system, thirteen rings altogether, the five most prominent being discovered in 1977 when astronomers observed a star that was being occulted by Uranus.

It is quite an organisational feat to make observations like this. The position of the star has to be measured with exquisite precision, and similarly accurate calculations made of where the planet will go. It might well be that, as seen from some places on the Earth, no occultation occurs, as the edge of the planet scrapes by, grazing the star but never actually covering it, not to mention that some potential observing stations might be in daylight at the critical moments, or covered by cloud.

One way around these logistical problems is to organise a number of observing stations to stand by, located in the right places. The campaign of 1977 took a different approach and used the Kuiper Airborne Observatory, a NASA facility built into a Lockheed C-141A Starlifter jet transport aircraft. It positioned itself high above the clouds in the right place at the right time, with the intention to study the fading of the star's light caused by its passage through the planet's atmosphere. This duly happened as expected, but the surprise was that the starlight was also unexpectedly dimmed five times before and five times afterwards. One group of five occurred forty minutes before, the other forty minutes after the main event. The individual dimmings in each group of five were of different depths, and the pattern of the depths was the same in the two groups, but the order of the 'after' group was reversed with respect to the 'before' group.

The reason for the extra dimming events was that there is a system of five rings that extend around Uranus. The rings are of different densities – that is why the dimmings had different depths. The rings were confirmed by images taken by the *Voyager 2* spacecraft, which flew through the Uranus system in 1986, and have been studied by the Hubble Space Telescope. They are confined and shepherded by Uranus's satellites. In some cases, the satellites are only hypothesised: they are small and have not yet been seen.

Uranus has one property unique among the main planets of the solar system: it has been bowled over. All the others spin like tops, with their spin axis almost perpendicular to their orbit around the Sun and rotating in the same sense that they orbit. As seen from the position of the North Celestial Pole, which lies in the stars immediately above the Earth's geographical North Pole, the Earth rotates anticlockwise and revolves anticlockwise. The Earth's equator is not tilted much compared to its orbital plane. Uranus, on the other hand, is almost topsy-turvy: it rolls along in its orbit with its spin axis lying in the orbital plane, in fact pointing a bit below the plane, a world turned upside down. The planet's satellites provided the first evidence of this unusual tilt. They orbit around Uranus's equator and show that its pole is tilted by more than 90 degrees. The rings are positioned likewise.

As a result, the north and south poles of Uranus point alternately, for half of Uranus's year of eighty-four

Earth-years at a time, towards and away from the Sun. At the midpoint of the planet's northern summer, its north pole points almost directly at the Sun. As Uranus moves on in its orbit, the Sun moves off centre, away from the north celestial pole. Someone on Uranus, near that pole, rotating once per 'day' on Uranus (it rotates once every seventeen hours) would see the Sun move in circles around the north celestial pole. It would always be daylight. The circles would gradually get larger, dipping lower towards the horizon each day, eventually skimming it. Twenty-one years after mid-summer, the Sun would fail to rise above the horizon. Forty-two years of winter would follow, a night-time as long as the earlier period of perpetual daylight, with the northern hemisphere perpetually dark and cold. Eventually, the Sun would peep above the horizon, and summer would return. The observer at the north pole would again experience uninterrupted sunshine.

An observer on the equator, by contrast, would see a day/night alternation every day, 8.5 hours of each. In mid-summer and mid-winter, the Sun would never rise far above the horizon, circling the north and the south celestial poles respectively. In the spring and autumn, the Sun would pass overhead each day. As a result of this cycle, the seasons of Uranus are much more extreme than ours. This may be connected with the sporadic appearance of the planet's methane storms, but the seasonal cycle of

Uranus is long (eighty-four years) and no one has seen enough of it up close to know how this might work.

What happened to tilt Uranus over so much? As in other cases where astronomers have had to explain some unique feature of the planets that took place a long time ago, there are more answers suggested than one – the secrets of planets' lives are often well hidden. One fact that might be significant is that the main satellites of Uranus and the inner rings orbit around its equator. Whatever caused the tilt of Uranus, at the same time caused the tilt of the satellites.

One theory is that at one time Uranus had a huge, close moon. This caused Uranus to wobble a lot as it spun. It wobbled onto its side, taking its satellites with it. Then the moon got ejected when Uranus encountered some other body in the solar system.

The theory that seems to be the most widely accepted, however, is that, at the end of the process by which Uranus formed, it was struck at an angle by a particularly large planetoid, Earth-sized or larger, that knocked it sideways before being absorbed into the planet. Uranus as a whole still 'remembers' this off-centre collision and its tilt is a consequence of the particular direction of attack from which the planetoid approached. The satellites would have been formed in part from the debris left over from the collision. The mess that is the planet's magnetic field could be the result of the collision, perhaps due to some funny structure that has been left in the planet's interior. A gloss on this theory is that Uranus was struck by two or more collisions, one after the other. However, none of the theories has been wholly successful in showing how all this happened and left the Uranus system in its present state.

The astronomy around Uranus has shown two completely different ways of approaching science. On the one hand, astronomy grew through astrology, a mystical pseudo-science based on arcane numerology, like Kepler's search for geometric or arithmetical formulae about the orbits of the planets. On the other, it also grew through meticulous and systematic observation, like Herschel's search of the heavens. The discussion of Bode's law lies somewhere between these two extremes, in a position that remains unresolved. Likewise, the solar system has two faces. On the one hand, it looks like an accurate watch, completely regular and orderly. On the other hand, it is also the result of chance and chaos, with catastrophic events that give each planet unique characteristics.

The lives of the planets are a mixture of orderly and accidental events. Our own lives consist of the same mixture, not only of events, but also of orderly, rational thoughts, and disorderly, irrational speculation.

CHAPTER 15

Neptune: the misfit

- ⊜ Scientific classification: *Ice giant.*
- ⊜ Distance from the Sun: *30.1 times the Earth–Sun distance, 4,495 million km (2,793 million miles).*
- ⊜ Orbital period: *165 years.*
- ⊜ Diameter: *3.88 times Earth, 49,528 km (30,775 miles).*
- ⊜ Rotation period: *19.1 days.*
- ⊜ Average temperature of the top of the clouds: *–200 °C.*
- ⊜ Secret complaint: *'Jupiter and Saturn ganged up to push me away, making me swap places with Uranus.'*

Neptune was discovered, a new world, at the tip of a pen. It was calculated to be where it was by a French mathematician, Urbain Le Verrier. It was found in the expected place, but it has turned out to be a misfit there: scientifically speaking, it is in the wrong place, thrown there by the chaos in the solar system.

Le Verrier had set out to solve the problem of why Uranus was going off course. After it had been discovered by William Herschel, astronomers were able to track down several previous observations of Uranus. They were observations in which it had been seen and recorded in catalogues and on star charts as a star by astronomers, who did not recognise it as a planet, because their telescopes were not clear enough. There were twenty-two occasions during the eighty-one-year period between December 1690 and December 1771 when Uranus was seen, before William Herschel discovered it by recognising it for what it was. By the 1820s its position in its orbit of eighty-four years had been measured well enough to be able to show that it was straying off course, and astronomers began to discuss why. One popular guess was that a previously unseen planet was pulling the planet off track.

Two mathematicians were attracted to the problem of calculating where the unseen planet was. They were Urbain Le Verrier in Paris and John Couch Adams in Cambridge. Adams' calculations were correct, but he was young and unassuming, and he was snubbed when he diffidently approached George Airy, the rather grumpy

Astronomer Royal, for some kind of assistance in looking for the new planet. To give him his due, Airy was the most senior scientist employed by the British government at the time, and was inundated with requests, many of them non-astronomical, such as investigating why a bridge fell down. Airy passed the buck to James Challis in Cambridge, who started a half-hearted search.

Challis and Adams were overtaken by events in 1846 when Le Verrier sent his prediction of the position of the unseen planet to Johann Galle, an astronomer at the Berlin Observatory. Galle, together with his assistant Heinrich D'Arrest, began a search on the same night that they received the letter, comparing some new star charts with the appearance of the region of the sky that Le Verrier had identified. Within thirty minutes, D'Arrest and Galle had identified a star that was not on the maps, and, on the following night, they confirmed that it was the new planet when they saw that it had moved. Galle wrote to Le Verrier, saying, 'Monsieur, the planet of which you indicated the position really exists.' Le Verrier replied, 'I thank you for the alacrity with which you applied my instructions. We are thereby, thanks to you, definitely in possession of a new world.'

When Holst composed *The Planets* between 1914 and 1916, Neptune was the outer boundary of the solar system. Perhaps that is why he composed this, the final movement, with a fade-out ending representing the indefinite spaces beyond the planet. The sound of its final chords, voiced by a women's choir in a room off stage, gradually falls away as the door of the room is closed, the final bar repeated until the sound is lost in the distance.

Losing its position as the world that marks the furthest reaches of the solar system, first to Pluto and then to the Trans-Neptunian Objects, Neptune nonetheless remains the outermost of the four giant planets. With Uranus it makes a pair, of the so-called 'ice giants'. Its atmosphere – the outer layer that we can see – is primarily hydrogen and helium. It has bands of weather systems like Jupiter and unexpectedly extensive storms, one of them an Earth-sized 'Great Dark Spot'. This storm is more prominent than the storms on Uranus, even though Uranus is nearer to the Sun and therefore receives more energy to drive its weather patterns. Although Neptune is more active than Uranus, that is not saying much – through a normal telescope it appears as a featureless pale blue globe. Its colour is bluer than Uranus's due to the greater amount of methane in its atmosphere (because it is colder). Curiously, Neptune emits more than twice as much energy as it receives from the Sun. The excess comes from the cooling of Neptune's hot interior. It has four rings, very faint and rather clumpy, perhaps fragmented asteroids or comets that passed too close and were captured and broken up by tidal forces. Neptune has been visited only once by a spacecraft, in a fly-by by *Voyager 2* in August 1989.

The frontier marked by Neptune before the discovery of Pluto and the Trans-Neptunian Objects is not a frontier where exciting things happen often. Because it is far from the Sun and the Sun's gravity is weak, things move slowly. It is so slow that Neptune could not have been formed here. How come?

There is a progression in the size of the planets outwards from the Sun – smaller ones near the Sun, massive ones in the outer zone, with masses tailing off towards the boundary of the solar system. This progression must have had its origins in the density of the solar nebula. The thought is that, at a given ring in the nebula around the Sun, the more material that was there in orbit, the greater the mass of the planet that would initially form there. Of course, there would be processes afterwards that would reduce or increase the mass of the planet from this time, or rearrange the planets. But what was the starting point?

The solar nebula is the disc of gas and dust out of which all of the planets formed, and of course it has now disappeared. We might be able to look at similar nebulae that orbit around nearby stars that are forming their own systems of planets, as a guide to what might have started our own solar system. But this is not practical, because even the nearest nascent planetary systems are too far away. Another problem is that astronomers can detect only gas and dust and can't see anything larger than a tennis ball unless it is the size of a planet or more. They can't detect the planetesimals that form within nebulae like this one. Because they are the essence of planet-building, these small bits are crucial.

As a result, astronomers have to address the problem of defining the beginning of the process by looking at its final outcome, namely our solar system, and working backwards. We know that the hydrogen and helium in the solar nebula was driven off many planets, but we can be fairly confident that the heavier elements in a planet, like iron and silicon, are representative of what it was born with. The idea, therefore, is to take the rocky component of each planet, and add hydrogen and helium until the chemical elements as a whole match the Sun in composition, on the assumption that the composition of the Sun has not changed much. Astronomers then spread that augmented mass for each planet over the area of its orbit to get a map of the surface density of the solar nebula. They then try to calculate how a nebula of that density would form planets.

Astronomers have found that this method, promising though it seemed to have been, does not succeed in making the planets of our solar system. The method gives low surface densities, with the mass of the solar nebula too spread out to form the giant planets quickly enough. Jupiter, according to this method, would take millions of years to form, Uranus and Neptune billions of years, whereas the indications are that the process took perhaps hundreds of thousands of years, or even less time.

It looks as if there has not been enough time for our solar system to develop. If the planets formed where they are now, it would take too long for enough material to fall together to make giant planets. Moreover, the longer the time that hydrogen and helium hang about, the more of it dissipates into space. The formation of the giant planets would not only slow down, but it would never be complete.

Rather than abandoning this line of attack, astronomers have looked at possible tweaks that they could give the theory to make it work. One promising tweak is to notice that, if the planets were formed at about halfway in from where they are now, this would compress the solar nebula into a quarter of the area, increase its density accordingly and begin the creation of planetesimals from a denser start, which would speed up the making of big planets.

This all suggests that the outer planets must have formed much closer to the Sun, and moved outwards. Through the mutual interaction of the planets, the solar system grew bigger. This is the essence of the Nice Simulation.

But there is one detail that is illuminating about Neptune. In the outer reaches of the solar nebula there was a smooth drop-off of surface density with distance from the Sun, that gives a smooth progression of planet mass – Jupiter is 320 times the mass of the Earth and Saturn 95 times. Then the progression goes wrong, throwing up something unexpected. Uranus is fourteen times the mass of the Earth, but Neptune is bigger at seventeen times. Neptune should be closer to the Sun than Uranus. On this argument, Uranus should be the frontier to the solar system, not Neptune. In other words, Neptune is in the wrong place.

To make the theory of planetary formation work out correctly, the two outer planets have to be switched around. Amazingly, this switch is something that happens in the Nice Simulation. Recall that the simulation begins with a variety of starting points for the numbers and positions of the planets, and then are run to see what happens. The outcomes are compared. Then everyone's attention focuses on the ones that best fit the real solar system. One feature of the good fits is that in half of them Uranus and Neptune swap over in their positions. This happens as a result of Jupiter and Saturn coming into resonance, with two orbits of the one fitting exactly into the time for one orbit of the other. The combined pull of the two biggest planets swapped Uranus and Neptune over.

In summary, Jupiter moved inwards at first but ended up moving outwards. Saturn, Uranus and Neptune also moved out, but, before it settled down into its current near-circular orbit, Neptune moved in an eccentric orbit in which it ranged over a large fraction of the solar system. It cut across the orbits of the other planets, jaywalking. It had a large effect on the smaller bodies of the solar system, the pieces that had not been eaten by the larger planets and that moved among them as asteroids. The combined effect of all this chaos was to throw many asteroids inwards, hurling some into trajectories where they became captured as satellites by some of the major planets, in the way, perhaps, that Phobos and Deimos became satellites of Mars (Chapter 7). Some asteroids became confined into the space near Ceres between the orbits of Mars and Jupiter (Chapter 8). Other asteroids were thrown outwards towards the edge of the solar system (Chapter 16). Yet others were scattered into the loneliness of interstellar space.

This was probably the most turbulent period in the lives of the planets. The outcome was, however, rather favourable for us. The asteroids that pervaded the solar system have been swept up and confined, as if the solar system was being tidied up, removing most of the risk that Earth would be bombarded in the future. Of course, we are still at risk from strays, so asteroids continue to affect the evolution of life on Earth, as the Chicxulub impact did for the dinosaurs. But impacts are occasional, rather than being a sustained and lethal bombardment. As a species we survive.

CHAPTER 16

Pluto: the outsider who came in from the cold

- Scientific classification: *Dwarf planet.*
- Distance from the Sun: *39.5 times the Earth–Sun distance, 5,906.4 million km (3,670 million miles).*
- Orbital period: *248 years.*
- Diameter: *0.186 times Earth, 2,370 km (1,464 miles).*
- Rotation period: *6.4 days.*
- Average surface temperature: *–225 °C.*
- Secret reflection: *'I was found as a planet and they were happy with me as a planet for seventy years (except for a few complainers). They don't want me as a planet any more, but I am happy now to lead an important new group.'*

Pluto was once classified as the ninth planet, revolving around the Sun in an orbit only a bit bigger than the orbit of Neptune. It certainly orbits the Sun and it is large enough, but after hanging around on the edges of the gang of the planets that rule the solar system, it proved not to be dominant enough, and was edged into a less influential mob of small worlds. It is more akin to the crowd of minor players known as Trans-Neptunian Objects than to the eight planets. Falling from grace, it was recently downgraded in status to 'dwarf planet'. Pluto is on the edge of the solar system, literally an outsider. Its change of status made it seem even more so, metaphorically, and some people protested at this further humiliation. It has turned out, though, that Pluto belongs to one of the most significant groups of worlds in our solar system.

Pluto had been discovered after a search intended to find the planet beyond Neptune. Neptune had been found by looking at the assumption that an unseen planet was pulling Uranus off course, and by the end of the nineteenth century Neptune itself was suspected likewise to be off course. Perhaps a ninth planet orbited outside the orbit of Neptune, a so-called Planet X. A businessman from Boston, Percival Lowell, who had established an observatory in Flagstaff, Arizona, took up the search and repeatedly photographed the sky to look for this planet.

Lowell had found no Planet X by the time of his death in 1916. The search lived on and was passed eventually to Clyde Tombaugh, a young amateur astronomer employed to take it forward. At the age of twenty-four, in 1930, Tombaugh discovered Planet X, later named Pluto. Found as a result of a search for a planet, it was at the time, naturally, identified as a planet.

Like Neptune, Pluto was found near to the place where it had been predicted to be. However, this was a matter of luck. The discrepancies in the orbit of Neptune proved to have been exaggerated, and Pluto is not very massive (its mass was not measured until the 1980s) and could not have caused them, so the predictions were meaningless: Lowell and Tombaugh were looking in the right place by chance. The search had borne fruit, but it was a completely different fruit from the one sought.

Although Pluto was acclaimed as Planet X, some oddities meant that it never really fitted in with the other planets of the solar system. It was curiously small, half the diameter of Mercury. It does have five satellites, one of them, Charon, of comparable size to Pluto and visible using ground-based telescopes, the others small and discernible from a distance only by the Hubble Space Telescope. But we now know that many of the smaller bodies of the solar system have satellites, and that this is not a property of particular importance.

Even before these facts had been uncovered and what was known of Pluto was principally its orbit, it was considered to be somewhat of an outsider among the planets. It goes around the Sun in an orbit that is very eccentric – it passes inside Neptune's orbit for part of the time. Also, the orbit lies at a surprisingly large, skewed angle to the orbital plane of all the other planets. These inconvenient differences between Pluto and the rest of the planets were set aside and, like Saturn, Uranus and Neptune successively before it in the history of astronomy, it took its place in textbooks as the outermost planet of the solar system, an outsider literally as well as, perhaps, metaphorically.

In a highly successful mission, the spacecraft *New Horizons* inspected Pluto from close up in a fly-by in 2015, after a ten-year journey from Earth. Pluto is cold, with a temperature averaging –230 °C (–380 °F). It has a rough, cratered terrain, with mountains of water ice and plains of frozen nitrogen, methane and carbon monoxide. The mountains are kilometres high. The landscape is similar to some of the rugged areas of Antarctica. Under a black, star-lit sky, its ice cover sparkles in what looks like bright moonlight: the light of the Sun is diminished on Pluto nearly to the brightness of the Full Moon here on Earth.

Pluto's largest moon, Charon, is huge in Pluto's sky. At 4 degrees, it is eight times the diameter of our own Moon but, because the Sun is so much further away from Pluto than the Earth, has only a few per cent of the brightness of our Moon, comparing the way Full Charon looks from Pluto with the Full Moon seen from Earth, and so on with the other phases.

Pluto has a tenuous atmosphere. As *New Horizons* left Pluto, it looked back at the atmosphere backlit by the Sun and saw a blue, layered haze of smog extending 200 kilometres (160 miles) high above the surface. The atmosphere is nitrogen with methane and other molecules, and the action of sunlight on these gases produces a mixture of hydrocarbons such as acetylene and ethylene. These chemicals accumulate into small particles, which cause the smog.

There is on Pluto a striking plain called the Sputnik Planitia, about 1,000 kilometres (600 miles) in dimension. It is made up of frozen nitrogen and carbon monoxide. It seems to be an impact basin, 3 kilometres (2 miles) deep, caused by the strike of a large meteor, which has become filled up with ices. Its surface is divided up into polygonal cells. They are thought to be circulating convection cells, with the ice churning up and around, driven from below by some source of energy, created by the original meteor impact, perhaps. Sputnik Planitia is akin to the lava-filled basins on the Moon, the grey *maria*, but made of ice rather than basaltic rock. Glaciers flow into the Planitia: glaciers of ice made of nitrogen, not water.

Pluto has many meteor craters – a thousand have been located on the part of Pluto that was surveyed in the fly-by. But Sputnik Planitia shows no meteor craters on its surface at all, so the impact that made it was recent – a catastrophe that happened less than a few million years ago. At its western edge, where the ice lies near the mountains, is a field of dunes, not of sand but of frozen methane particles. The particles have been blown into dunes by the wind that is associated with the mountains. It was a surprise to scientists that methane snowflakes could be mobilised by the wind on Pluto, because its atmosphere is so tenuous, and Pluto is so far from the Sun that the heat available to generate motions in the atmosphere is minimal. However, after the calculations had been done, they showed that the surprise was founded in lack of imagination about the conditions on other worlds; there was no new underlying science.

In 1992, David Jewitt and his PhD student Jane Luu discovered the first of over a thousand more small planets outside the orbit of Neptune. They have the descriptive but unexciting name of Trans-Neptunian Objects (TNOs). There are probably hundreds of thousands of them larger than 100 kilometres in size. Some of them are as large as or even larger than Pluto. Most of them occupy a zone called the Kuiper Belt at the outer edge of the solar system.

Sunlight is weak out there, TNOs are a long way from Earth, and most are small; they are rather dim and difficult to see, so they successfully hide from us.

TNOs were mostly made in the region where they are now, and some were ejected into the Kuiper Belt from closer into the Sun, perhaps by the resonance of Jupiter and Saturn as described in the Nice Simulation. They are thought to be, in the main, unchanged planetesimals surviving from the early days of the solar system. Some of them proved to be of considerable size, comparable with Pluto. A lot of the TNOs were discovered through telescopes on Hawaii and given names from Hawaiian culture. Haumea and Makemake are two of the bigger ones.

Ultima Thule is one of the small, faint TNOs. It was discovered by the Hubble Space Telescope in a deep search in a very particular region of the solar system, namely the region through which *New Horizons* would fly after leaving Pluto. The HST search aimed to provide a target in the Kuiper Belt for the spacecraft to investigate. It found three and Ultima Thule was the TNO chosen for study.

Ultima Thule orbits with a period of 298 years at a distance 44.5 times the distance of the Earth from the Sun. It was given the name Ultima Thule in a public competition. The name refers to the legendary island

Thule, said by the Vikings to lie north of Britain. The island Thule was thought in medieval times to be the furthest land to the north and was so described as 'ultima' ('furthermost'). The eponymous TNO Thule is not in fact the furthermost TNO. A TNO about five hundred kilometres in diameter was discovered at the end of 2018 and currently lies about one hundred and twenty-five times further from the Sun than the Earth, three times more distant that Ultima Thule. The new TNO has a nickname, not yet accepted as official: Farout. We will soon run out of descriptive names for these distant objects, as astronomers find more and more of them, further and further away.

The TNO Ultima Thule was visited by *New Horizons* on the first day of 2019 and proved to be shaped as two lobes touching each other, from some angles looking much like a snowman. Its overall length is 33 kilometres (20 miles). The lobes are icy and lumpy, with holes that might be craters but might also have been excavated by the escape of gases. The lumps are smaller planetesimals that have stuck together to make each sphere separately, with the two lobes then gently colliding and fusing together. These two parts of Ultima Thule are among the most primitive of the solar system, first made 4 billion years ago or more, and unchanged since then, except for the collision.

If Ultima Thule was much bigger, perhaps more than ten to a hundred times the size that it actually is, its gravity and the warmth generated by radioactivity in its interior would have caused it to differentiate into layers and settle into a spherical shape. This is what happened to Pluto. However, the shape of Ultima Thule still remembers the TNO's earlier life.

With the fresh insight brought by all these discoveries in the Kuiper Belt, the status of Pluto becomes clearer. It is no longer seen to be unique in the solar system: there are other objects like it in similar, tilted, eccentric orbits, near and beyond Neptune. Pluto is not really a planet, it is a TNO.

The proposition that Pluto is not a planet caused considerable public discussion in the first years of the new millennium. It is not easy to say why the general public, particularly in the USA, became so concerned about this scientific issue. What is clear is that there must have been emotional reasons. Pluto was the only planet that had been discovered by an American – was that the reason? Or was it that, now that Greek mythology is less prominent, the name Pluto is quite cuddly, having been shared with the Disney cartoon dog, Mickey Mouse's pet. (The dog was, according to Disney family tradition, named after the planet, soon after its discovery.)

In the way that science is normally practised, an issue about the nature of Pluto would not have been decided on a specific occasion, like a motion before parliament. As happened earlier in the case of the asteroids, the issue would have been discussed by individual astronomers on many occasions, with the subtleties being teased out one by one. The astronomers most concerned would have expressed different shades of opinion, as Herschel, Piazzi and Bode did about asteroids (Chapter 8). Other astronomers would have followed the argument that they found most convincing. Some of them would have summed up the issue in lectures, articles or textbooks. The consensus would have emerged gradually. However, this did not happen in the argument over Pluto. The argument became political, and found its way to the International Astronomical Union (IAU), which came to agreement in a way that is unusual in science.

The IAU is an organisation that brings the astronomers of the world together the better to co-ordinate their work. It agrees naming conventions in order to give some commonality to the cataloguing of celestial objects, so as to facilitate the drawing together by scientists of various items of data. The labelling of astronomical objects is important to the IAU so that it can draw up its lists.

The IAU meets every few years in a General Assembly where the latest issues in astronomy are discussed. It met in Prague in August 2006, where the Pluto question was discussed a lot. After much preliminary skirmishing, the proposal was put to a formal meeting on the last day. It passed by a large majority.

The IAU agreed a list of properties that define a planet. A planet, it said, is a celestial body orbiting the Sun that is large enough to make itself close to spherical (unlike comets, or nearly all asteroids, with Ceres the notable exception; see Chapter 8). For this part of the definition of a planet, size matters, and planets would have to be more than about 400 kilometres (about 300 miles) in diameter, perhaps considerably more, depending whether they are made of rock or ice. An additional part of the definition of a planet, according to the IAU, is that it is also a celestial object large enough to cause all other objects (except for any satellites they might have) eventually to leave it isolated in its orbit, either absorbing the extraneous objects or ejecting them.

I was one of the attendees who voted to approve the proposal. Although I had reservations about the process, I was motivated to vote by wanting to see an end to the argument. The process had been long drawn out, and there were aspects that were intractable. But the problem needed to be solved and put aside – we astronomers were looking foolish, like medieval theologians arguing about the degrees of angels.

The IAU's definition of 'planet' combines three separate criteria, each of a different nature. The first criterion, about a planet's orbit, is what, since Copernicus, everybody had thought was the main property of a planet. The second criterion is about the structure of a planet – a planet is massive enough to settle down into a spherical world, and balance its internal structure under its own force of gravity. This is also what we thought, the reason why planets look the way we expect: approximately spherical. Pluto passes these two tests.

The last criterion is that at the end of its birth process a planet will have 'cleared the neighbourhood' of its own orbital zone, either absorbing or ejecting other bodies of comparable size (other than its own satellites). Pluto has not done this, because it orbits in company with other TNOs. So, Pluto is not a planet.

If Pluto is not a planet, what is it? The IAU defined a second category of solar system bodies, namely 'dwarf planets':

A 'dwarf planet' is a celestial body that (a) is in orbit around the Sun, (b) has sufficient mass for its self-gravity to overcome rigid body forces so that it assumes a hydrostatic equilibrium (nearly round) shape, (c) has not cleared the neighbourhood around its orbit, and (d) is not a satellite.

The IAU noted that Pluto is a 'dwarf planet' by these definitions, like the asteroid Ceres (Chapter 8). Pluto had been demoted from 'planet' to a 'dwarf planet'. However, if its pride was hurt, it could take consolation from the fact that it led the group of dwarf planets in the solar system as a typical example: they include Ceres, Haumea, Eris and Makemake, but there are many TNOs that are likely dwarf planets, and there may be many more large dwarf planets so far completely undiscovered in the darker reaches of the solar system.

Pluto may have been demoted, in a sense, but it is no longer an outsider, either literally or metaphorically. It may lie near the edge of the solar system, it may be literally icy, and many details of its biography may still be unknown. But it is one of an important group of objects in the solar system that hold the key to the birth of the planets. We have recognised Pluto for what it is, and we have learnt some secrets of its life. It has been welcomed in from the cold.

The solar system in a nutshell

THE SUN

The Sun is the star at the centre of the solar system. It is by far the largest body in the solar system and, to a good approximation, all the other bodies revolve around it, although it is a bit more accurate to say that they all, including the Sun, revolve around their common centre of mass. The Sun generates its own energy, whereas all the other bodies in the solar system radiate reflected sunlight, boosted a little by energy radiated because of their warmth.

THE PLANETS

Planets are the large bodies – solid, liquid and/or gaseous – that revolve around the Sun. As a result of their large mass they have settled into a nearly spherical shape, built up layer by layer, with each layer supporting the weight of the layers above. They have also gathered up all the material that lay near to their orbit at the time of the formation of the solar system by feeding on it, with the exception of bodies called **satellites** or **moons**, which orbit around planets. The planets are Mercury, Venus, Earth, Mars, Jupiter, Saturn, Uranus and Neptune. The first four have a rocky surface and are called **terrestrial planets**, the last four have an extended gaseous envelope and are called **gas giant planets**, with the last two also sometimes called **ice giant planets**.

DWARF PLANETS

Planets that orbit the Sun and are spherical, but share an orbital zone with other similar bodies, are called **dwarf planets**. The dwarf planets include Ceres, which orbits between Mars and Jupiter, and Pluto, Haumea, Makemake and Eris, which orbit in the outer solar system. There are about a hundred likely dwarf planets in the solar system, but there may be hundreds more, as yet undiscovered.

Everything orbiting the Sun that is not a planet, a dwarf planet or a satellite is called a **small solar system body**. This category includes:

ASTEROIDS

Minor planets that, in the main, orbit between Mars and Jupiter in the **Asteroid Belt** are called **asteroids**. Some asteroids range in orbits elsewhere in the solar system. They are all rocky. One is spherical, Ceres, but the rest are substantially non-spherical, and support themselves, not layer by layer, but as solid pieces.

METEOROIDS

Meteoroids are rocks or dust particles, very similar to asteroids but smaller.

METEORS

Meteors are meteoroids that are in the act of falling onto a planet, asteroid or satellite, perhaps burning up in an atmosphere, as in the meteors we see flashing through the night sky on Earth.

METEORITES

Meteorites are meteoroids that have fallen to the surface of a planet, like on Earth.

TRANS-NEPTUNIAN OBJECTS

Trans-Neptunian Objects (TNOs) are minor planets that orbit beyond Neptune in the **Kuiper Belt**, and include Pluto and all the dwarf planets except Ceres.

COMETS

Comets are small solar system bodies made of ice and rocks, which orbit anywhere in the solar system and which, when they approach the warmth of the Sun, emit vapour and dust in a hazy cloud and as a trailing tail.

Timeline

6,000 to 9,000 BCE The Ishango bone is engraved with a record of the lunar phases.

500 BCE Pythagoras points out that the two manifestations of Venus are one planet.

1543 Nicolaus Copernicus publishes his theory that the Sun is the centre of the planetary system.

1609–19 Johannes Kepler discovers the mathematical properties of the orbits of the planets.

1610–16 Galileo Galilei turns a telescope to the sky for the first time and, among other things, discovers the phases of Venus, the moons of Jupiter, and the rings of Saturn.

1656 Christiaan Huygens identifies the changing appearance of Saturn as caused by its rings.

1665 Giovanni Cassini discovers Jupiter's Great Red Spot.

1666 Giovanni Cassini discovers polar caps on Mars.

1675 Giovanni Cassini discovers the internal structure of Saturn's rings.

1686 Bernard de Fontenelle publishes his book *Conversations on the question whether there are other worlds.*

1687 Isaac Newton puts forward his theory of gravitation and explains the motion of the planets.

1766 Johann Daniel Titius introduces his formulation of the Titius–Bode law into his textbook.

1772 Johann Elert Bode publishes the Titius–Bode law.

1774 Nevil Maskelyne weighs the Earth at Schiehallion.

1781 William Herschel, with his sister Caroline, discovers Uranus.

1796 Pierre Simon Laplace demonstrates that the solar system is stable.

1801 Giuseppe Piazzi discovers the first asteroid, Ceres.

1802 Wilhelm Olbers discovers Pallas.

1802 William Paley describes the construction of the planetary system as being like a watch.

1804 Karl Ludwig Harding discovers Juno.

1807 Wilhelm Olbers finds his second asteroid, Vesta.

1815 The Chassigny Martian meteorite falls to Earth.

1827 Joseph Fourier discovers the greenhouse effect in the Earth's atmosphere.

1840 Wilhelm Beer and Johann von Mädler make the first maps of Mars.

1846 Urbain Le Verrier predicts the position of Neptune.

1846 Johann Galle and Heinrich D'Arrest discover Neptune.

1848 Édouard Roche shows how tidal forces can disrupt a planet's close satellite.

1859 Charles Darwin puts forward his theory of evolution by natural selection.

1859 Urbain Le Verrier initiates the search for Vulcan.

1860 Emmanuel Liais suggests that Mars's markings are patches of vegetation.

1865 The Shergotty Martian meteorite falls to Earth.

1877 Asaph Hall discovers that Mars has two satellites, Phobos and Deimos.

1877 Giovanni Schiaparelli maps Mars and claims to see canals.

1887 Henri Poincaré studies the Three-Body Problem and discovers chaos.

1894 Percival Lowell sets up the Flagstaff Observatory to monitor Mars and find Planet X.

1896 H.G. Wells writes *War of the Worlds*.

1909 Eugenios Antoniadi shows that the Martian canals are illusory.

1911 The Nakhla Martian meteorite falls to Earth.

1913 Milutin Milankovič calculates the cyclic variations of the Earth's orbit and the weather.

1915 Albert Einstein discovers General Relativity.

1930 Clyde Tombaugh discovers Planet X, later named Pluto.

1935 Prediction of the existence of metallic hydrogen by Eugene Wigner and Hillard Bell Huntington.

1936 Inge Lehmann uncovers the structure of the Earth's core.

1956 Measurement of the high temperature of Venus by Cornell H. Mayer.

1961 Carl Sagan explains the high temperature of Venus as the result of the greenhouse effect.

1962 *Mariner 2* becomes the first space visitor to Venus.

1963 Edward Lorenz discovers chaos in weather forecasting.

1965 *Mariner 4* is the first successful mission to visit Mars.

1969–72 *Apollo* astronauts land on the Moon.

1970–6 *Luna 16, 20* and *24* return lunar soil to Earth.

1970–83 *Venera* missions to Venus, including *Venera 7*, the first probe to land on the surface of another planet.

1971 *Mariner 9* is the first spacecraft to enter into orbit around another planet, Mars.

1974–5 *Mariner 10* is at Venus and then Mercury.

1975–6 *Viking* missions land on Mars.

1977 Discovery of the rings of Uranus by the Kuiper Airborne Observatory.

1978 Glen Penfield discovers the Chicxulub crater.

1979 Linda Morabito discovers volcanoes on Io.

1979 Saturn and its satellites explored by *Pioneer 11* fly-by.

1979 *Voyager 1* and *2* fly by Jupiter.

1980–1 *Voyager* spacecraft explore Saturn and its satellites.

1986 *Voyager 2* visits Uranus.

1989 *Voyager 2* visits Neptune.

1990 Mark Showalter discovers Saturn's satellite Pan.

1990–4 *Magellan* mission to Venus.

1992 Dave Jewitt and Jane Luu discover the first Trans-Neptunian Object after Pluto.

1994 Comet Shoemaker-Levy 9 plunges into Jupiter's atmosphere.

1995 *Galileo* becomes the first spacecraft to enter into orbit around Jupiter.

2000 *NEAR Shoemaker* enters into orbit around Eros, landing in 2001.

2003 *Beagle 2* taken to Mars on *Mars Express*.

2004 *Cassini-Huygens* spacecraft explores Saturn.

2004 The *Huygens* probe parachutes onto Saturn's largest satellite, Titan.

2005 Alessandro Morbidelli and collaborators publish the Nice Simulation.

2005 *Cassini* discovers geysers on Enceladus.

2005 *Hayabusa* probe studies Itokawa.

2006 IAU creates the modern definition of planet.

2009–18 *Lunar Reconnaissance Orbiter* explores the Moon.

2011–12 *Dawn* spacecraft orbits Vesta.

2011–15 *Messenger* orbits Venus.

2013 *Chang'e* space mission takes its lander *Yutu* to the Moon.

2015 *New Horizons* explores Pluto.

2015 *Dawn* spacecraft orbits Ceres.

2016 *Juno* enters orbit around Jupiter.

2016 *OSIRIS-Rex* launched towards Bennu.

2018 *BepiColombo* launched.

2018 *Hayabusa2* probe studies Ryugu.

2019 *New Horizons* flies past Ultima Thule.

Picture Acknowledgements

1. Caloris Basin: © NASA/Johns Hopkins University Applied Physics Laboratory/Carnegie Institution of Washington
2. Venus: ©NASA/JPL
3. Earth: ©NASA/JSC/Apollo 17
4. Mars: ©ESA/DLR/FU Berlin
5. Mars sand dunes:©NASA/JPL/University of Arizona
6. Phobos: ©NASA/JPL-Caltech/University of Arizona
7. Ceres: ©NASA/JPL-Caltech/UCLA/MPS/DLR/IDA
8. Jupiter: ©NASA/JPL-Caltech/SwRI/MSSS/ Gerald Eichstädt /Seán Doran
9. Io: ©NASA/JPL/University of Arizona
10. Europa: ©NASA/JPL-Caltech/SETI Institute
11. Saturn: ©NASA/JPL/Space Science Institute
12. Pan: ©NASA/JPL-Caltech/Space Science Institute

13. Saturn Enceladus... ©NASA/JPL/Space Science Institute
14. Saturn Titan: ©ESA/NASA/JPL/University of Arizona; processed by Andrey Pivovarov
15. Pluto: ©NASA/JHUAPL/SwRI

Index

Index

Index

Index